Geoheritage, Geoparks and Geotourism

Conservation and Management Series

Spectacular geo-morphological landscapes and regions with special geological features or mining sites, are becoming increasingly recognized as critical areas to protect and conserve for the unique geoscientific aspects they represent and as places to enjoy and learn about the science and history of our planet. More and more national and international stakeholders are engaged in projects related to "Geoheritage", "Geo-conservation", "Geoparks" and "Geotourism" and are positively influencing the general perception of modern Earth sciences. Most notably, "Geoparks", have proven to be excellent tools to educate the public about "Earth Sciences". And shown to be areas for recreation and significant sustainable economic development through geotourism. In order to develop further the understanding of earth sciences in general and to elucidate the importance of earth sciences for Society the Geoheritage, Geoparks and Geotourism Conservation and Management Series has been launched together with its sister GeoGuides series. "Projects" developed in partnership with UNESCO, World Heritage and Global Geoparks Networks, IUGS and IGU, as well as with the "Earth Science Matters" Foundation, are welcome. The series aims to provide a place for in-depth presentations of developmental and management issues related to Geoheritage and Geotourism as well existing and potential Geoparks. Individually authored monographs as well as edited volumes and conference proceedings are welcome in this series. This book series is considered to be complementary to the Springer-Journal "Geoheritage".

More information about this series at http://www.springer.com/series/11639

Joan Martí · Llorenç Planagumà
Editors

La Garrotxa Volcanic Field of Northeast Spain

Case Study of Sustainable Volcanic Landscape Management

Springer

Editors
Joan Martí
Institute of Earth Sciences Jaume Almera
CSIC
Barcelona
Spain

Llorenç Planagumà
Tosca Environmental Services
Olot
Spain

ISSN 2363-765X ISSN 2363-7668 (electronic)
Geoheritage, Geoparks and Geotourism
ISBN 978-3-319-42078-3 ISBN 978-3-319-42080-6 (eBook)
DOI 10.1007/978-3-319-42080-6

Library of Congress Control Number: 2016948277

Printed on acid-free paper

This Springer imprint is published by Springer Nature
The registered company is Springer International Publishing AG
The registered company address is: Gewerbestrasse 11, 6330 Cham, Switzerland

Preface

The Neogene-Quaternary extension-related alkaline volcanism represents one of the most significant magmatic events in Western and Central Europe since the Variscan orogeny. Primitive mafic alkaline volcanic rocks are widely distributed along an extensive rift system which comprises among others the volcanic areas of the Rhenohercynian of Germany, the Bohemian massif in Czechia, the Massif Central of France and the Catalan Volcanic Zone in the NE of Spain. From this last the La Garrotxa Volcanic Field is the best example of a volcanic landscape in the Iberian Peninsula and one of the most interesting and best preserved in continental Europe. It formed from about 700 ka to early Holocene and has around fifty well-preserved cones and more than 20 basaltic lava flows with special morphologies. This volcanic field is composed of monogenetic cones, generated during Strombolian and phreatomagmatic eruptions, which formed along tectonic fissures and faults and include a large variety of successions of deposits. After some eruptions, lava emissions blocked the drainage of some valleys and created dams. These old lakes have been covered by sediments and nowadays are fertile plains.

The industrial and urban growth experienced in the region during the 1970s led to a series of serious attacks, including a massive extraction of volcanic materials, which seriously threaten all of its natural values. This impelled different sectors of society to mobilise leaving them in 1976 to create the Promotion Committee for the Protection of Volcanic Zone and the celebration, a year later, of the Campaign for the Protection of the Natural Heritage of the Catalan Countries, organised by the Congress of Catalan Culture. Finally, the Parliament of Catalonia unanimously approved the Law 2/1982, of March 3, the protection of the volcanic zone of La Garrotxa, declaring part of it as Natural Area of National Interest, with the aim of attending the conservation of its flora, geomorphology and of its special beauty, given the uniqueness characteristics of the territory (Art. 1). This new law also declared twenty natural reserves of geo-botanical interest, in order to avoid any action that could bring to their destruction, impairment, disfigurement or the transformation of its geomorphology or flora (Art. 2). The Decree 71/1986 of 13 February, approved the specific topographic boundaries of the natural park and the natural reserves included, and describes the external perimeter of the park and the cadastral sectors included in the nature reserve. The souther sector of La Garrotxa Volcanic Field was not included in the limits of the natural park but has also received attention from the local society and administration resulting in the declaration of several natural sites with special protection as is the case of the Crosa de Sant Dalmai volcano. All this new legislation permitted to stop mining activities and to minimise and restore the ruined geological heritage through the structuring and consolidation of the natural park and other protected zones. Example of such activities is the restoration in 1995 of the Croscat volcano, the most emblematic in the park, the youngest in the Iberian Peninsula and the one that has suffered most environmental impact.

The region occupied by La Garrotxa Volcanic Field is characterised by an intimate relationship between volcanoes and society, in an extent comparable or even superior than in areas with more active volcanism. In this region volcanoes are present in many aspects of local society, as its cultural heritage, local history, architecture, or even in its excellent cuisine.

People are aware of living among volcanoes and that they represent the most characteristic feature of their region.

This is a land rich in traditions and history, with some Romanesque gems, medieval villages (Santa Pau, Besalú) and deep-rooted popular festivals, in which a sustainable tourism is promoted through an excellent network of hotels, guesthouses, farmhouses, hostels, rural apartments and camping sites and with the main aim of showing to the visitors the volcanoes and their related landscape.

La Garrotxa Volcanic Zone Natural Park has been pioneering in many initiatives addressed to preserve their landscape and natural values and to promote their knowledge among the society. This book seeks to explain all these aspects to the reader in a pleasant and enjoyable way through the La Garrotxa volcanoes. In addition to a general description of the main geological and volcanological values of the region we also include a detailed description of the history of the region, its biological diversity, and its cultural heritage including architecture, folklore and gastronomy. An important part of the book will be also dedicated to describe the educational programmes and outreach developed to disseminate the main values of this region, as well as how a sustainable tourism has been implemented and the management plan that has been designed to preserve such important natural and cultural values.

The book is written by different local experts on the different topics covered in its chapters and is addressed to a general audience interested in visiting the area but also in knowing an example of geoheritage and geoconservation with a successful integration of education, tourism, planning and environmental management. Through the different chapters of the book the reader will get a detailed view the main natural and social values of La Garrotxa Volcanic Field.

The Book Contents

The chapters included in this book provide a general but also detailed overview of the main aspects that characterise La Garrotxa Volcanic Field and that make it a most singular protected area for what concerns the integration of natural geoheritage with society. Thus, through the different chapters of the book we will learn about the geology, landscape, habitats, history and culture of this exceptional zone, as well as on the management and educational programmes that have been made possible its conservation and the dissemination of their main values among the general public.

Chapter 1 introduces the main characteristics of the volcanic landscape of La Garrotxa describing it as the dynamic combination of a series of geological, biotic and anthropic elements that have converted this area into one of the most singular scenic areas of volcanic origin. The chapter revises in detail all these aspects and provides also an introduction on the European and Catalan legal provisions regarding landscape protection, planning and management, with special emphasis on the Catalan landscape catalogues and maps, which helps to better understand the singularity of this volcanic area.

Chapter 2 describes the main geological features of the area and offers the reader an overview of the characteristics of La Garrotxa volcanism and its current scientific knowledge. It lets the reader know how the eruption sequences differ from one volcano to another, combining episodes of lava flow emission, with others of poorly explosive magmatic (strombolian) activity or with highly explosive episodes (phreatomagmatic) resulting from the interaction of the erupting magma with groundwater, and how these different eruption behaviours depend on the differing stratigraphic, structural and hydrogeological characteristics of the substrata below each volcano.

Chapter 3 helps to understand how relief, diversity of substrata, climate and its geographical position in the eastern Pyrenees, has converted La Garrotxa into a region with a fascinating biological history, characterised by a great diversity of species and habitats. It also reviews how intense anthropic disturbance has also played its part and agriculture, animal husbandry, forestry and industry have helped increase the diversity of environments and afforded greater importance to the secondary communities and open spaces present in this region.

Chapter 4 shows how La Garrotxa is a land through which a multitude of different peoples have passed and left their mark, leaving behind a reach history in which the different periods are clearly identified by their imprinting on the human and cultural evolution of the region.

Chapter 5 complements in part to Chap. 2 and reviews the main features of the volcanism in La Garrotxa Volcanic Field trough the description of a number of selected outcrops and viewpoints of this volcanic field, also indicating how to reach them, with the aim to help readers interpret the landscape and the volcanological processes that characterise this volcanic field. Additional information on landscape, natural habitats and cultural heritage are also provided in cases where such information is particularly relevant.

Chapter 6 describes the main sites of natural and cultural or historical interest that can be visited in La Garrotxa Volcanic Field, in addition to the volcanological sites described in the previous chapter.

Chapter 7 explains the reader how the pedagogical activities carried out in this protected area have been one of the pillars to build on it a solid culture of preservation of the natural heritage. It describes how the use of the environment as a setting and as an educational resource, among other aspects, has helped to ensure the conservation of the region's geological, botanical and scenic riches.

Chapter 8 shows how in La Garrotxa sustainable tourism is being promoted via the European Charter for Sustainable Tourism, whose principles and strategies aim to promote local sustainable development, and revises what has been the economic impact of this sustainable tourism on the region.

Finally, Chap. 9 reviews the management of the geological heritage of La Garrotxa Volcanic Zone Natural Park, and includes both an evaluation of the strategy that was approved in 2004 and a discussion of the current and future work devoted to conserving the area's remarkable natural heritage and fomenting its sustainable use.

Barcelona, Spain Joan Martí
Olot, Spain Llorenç Planagumà

Acknowledgements

Writing this book has been possible, thanks to the information and support provided by several institutions and individuals. In particular, we want to thank the Documentation Centre of the Natural Park of La Garrtoxa Volcanic Zone for the material provided and for authorisation to reproduce part of it in the book. We also thank the staff members of the Natural Park for all technical queries resolved by the Park. Our gratitude also goes to the Olot Volcanoes Museum and to Tourism Garrotxa for providing information and graphic documentation that is included in the book. Eduard Masdeu has provided some of his nice aerial views of La Garrotxa to be reproduced in Chap. 2. Finally, we also thank the VUELCO project (European Commission FP7 Theme: ENV.2011.1.3.3-1; Grant 282759), which has permitted to undertake the most recent research on the geology and volcanology of the area.

Contents

The Character of the Volcanic Landscape of La Garrotxa

1

Joan Nogué and Pere Sala i Martí

All landscapes are the perceptible product of a dynamic combination of abiotic (geological substrata), biotic (flora and fauna) and anthropic (human activities) elements that convert the whole into a social and cultural amalgam in a continual state of evolution. The landscape is thus both a physical reality and the cultural representation that we make of it. It is the visible external physiognomy of a certain part of the Earth's surface and the individual and social perception that it generates—a tangible geographical entity whose interpretation is intangible. It is both the signifier and the signified, the container and the contents, reality and fiction. That is, it has a physical, material dimension *and* a spiritual, symbolic and perceptive dimension. Human societies use their cultures to transform original landscapes into cultural landscapes that are characterised not only by certain types of constructions, for example, but also by the translation of human values and sentiments onto the very landscape. Thus, the volcanic landscape of La Garrotxa is a paradigm as it encompasses and concentrates all these ingredients in a highly singular fashion that render it attractive on both a European and world scale.

One of the most singular scenic areas in the whole of Catalonia is the volcanic landscape of La Garrotxa, a montane landscape articulated by the rivers Fluvià and Ser that possesses a series of outstanding morphological features generated by the area's Quaternary vulcanism (e.g. the basaltic cliff at Castellfollit de la Roca and the 40 catalogued volcanoes, all of which are protected by La Garrotxa Volcanic Zone Natural Park, the entity in charge of the protection of the main elements of natural interest in this geological landscape). The volcanic landscape of La Garrotxa is also characterised by a combination of thick holm and deciduous oak and beech forests (including the singular D'en Jordà beech forest), and productive agricultural areas such as the Vall d'en Bas (located at the foot of the mountain of Puigsacalm) and the Vall de Bianya. Amidst this mosaic of fields, pastures, woodlands and farmsteads appear medieval villages such as Santa Pau and Besalú, and the urban landscape of the city of Olot, which is home to features of great urban interest such as the Malagrida New Town and Parc Nou.

Nevertheless, before examining more closely the main features of this landscape, it is worth considering briefly the European and Catalan legal provisions regarding landscape protection, planning and management, with special emphasis on the Catalan landscape catalogues and map.

Since 2000, both international (European Landscape Convention) and national (law aimed at protecting, managing and organising the landscape of Catalonia) directives have provided tools and instruments that can be used as forthright means of improving landscapes and strengthening their personalities.

1.1 The Catalan Landscapes

The European Landscape Convention: defending the enormous diversity and richness of European landscapes. The main point of reference in terms of the management and planning of European landscapes is the European Landscape Convention (ELC), adopted in Florence in 2000. This document describes the steps needed to be taken in European countries and regions to foment the conservation of and improvements in the extraordinary variety and richness of European landscapes, above all the most degraded. The Convention recognises that an attractive, appealing and harmonious environment generates a sense of well-being that significantly increases people's quality of life.

J. Nogué (✉)
University of Girona (UdG), and Catalan Landscape Observatory, Olot, Spain
e-mail: joan.nogue@catpaisatge.net

P. Sala i Martí
Catalan Landscape Observatory, Olot, Spain
e-mail: pere.sala@catpaisatge.net

J. Martí and L. Planagumà (eds.), *La Garrotxa Volcanic Field of Northeast Spain*, Geoheritage, Geoparks and Geotourism, DOI 10.1007/978-3-319-42080-6_1

The approval of the ELC unleashed a veritable revolution in the way in which landscape policy was decided upon throughout Europe. According to the ELC, all landscapes are important—urban and rural, pristine and degraded, part of protected areas or more mundane 'everyday' landscapes—and in all these spaces the landscape represents an essential element of individual and social well-being that contributes to the consolidation of the collective identity.

The ELC requires of European governments a series of responses.[1] It is worth highlighting just two: firstly, the landscape has become—for the first time—a fundamental criterion to be taken into account in territorial and urban planning, as well as in any cultural, agricultural, economic or other activity that could have a direct or indirect incidence on the landscape; secondly, the governments that have ratified the Convention now have the obligation to adopt procedures that will guarantee social and institutional participation in the application of landscape policies.

The great virtue of the ELC is that it is a meeting point for views of a concept—landscape—on which, given the enormous cultural significance referred to above, it is not easy to reach a consensus. The appeal of the ELC is that it is fruit of a continent-wide agreement on the question of landscape and, such is its relevance, it is now difficult to conceive any research, teaching, professional, political or legislative project affecting the landscape that has not been inspired by this document.

How has the ELC affected Catalonia? In December 2000, just two months after its approval, the Catalan Parliament signed the ELC, a first step that reflects this institution's interest in the conservation and improvement of the landscape. Nevertheless, the most fundamental steps were taken first at the end of 2004, with the creation of the Catalan Landscape Observatory, and then in the following year, with the passing of Law 8/2005, 8 June, for the protection, management and organisation of the landscape.[2]

The Catalan Landscape Law was passed with the aim of making economic and urban development compatible with the maintenance of the quality of the environment via a greater awareness of heritage and cultural, social and economic values. It also contained a fitting emphasis on the importance of recognising our landscape for what it is, knowing how to enjoy it and encouraging people to participate in the decisions that will affect it in the future. Taking the Law as a baseline, the Catalan government (Generalitat de Catalunya) has drawn up landscape catalogues (useful tools for landscape organisation and management in territorial planning) and landscape 'cards' (local or county-wide strategies that involve local administrations and social and economic actors), as well as providing funding (along with other administrations, entities and private enterprise) for protecting, managing and organising the landscape via projects in urban, peri-urban and rural areas. The government has also drawn up protocols for incorporating landscape criteria into the planning and execution of projects. Nevertheless, it is still too early to talk about a genuine policy on the landscape since the challenge of increasing awareness and improving education amongst the general public has still to be tackled, a task that should be carried out in a progressive fashion.

The first result of the Catalan Law on the Landscape was the setting up (1 March 2006) of the Catalan Landscape Observatory. This body was conceived as a tool for assessing the Catalan Government on landscape-related questions, raising awareness in general vis-à-vis the landscape in Catalonia, and working on the application of the ELC in Catalonia. The Observatory is a centre of thought, study, documentation and action on the landscape, which also encourages ideas and projects in collaboration with their promoters. Since its beginnings in 2006, it has become a forum for the Catalan government, local administrations, universities, professional bodies and Catalan society in general for debating all matters pertaining to the landscape.

One of the main tasks of the Observatory is the drawing up of the landscape catalogues of Catalonia. These documents have three principal aims: (1) use improved knowledge of the landscape when implementing landscape protection, management and organisation in territorial, urban and sectorial planning; (2) integrate new uses and constructions into the character of every landscape; and (3) foment a type of landscape governance based on participation and consensus. The first results of these tasks were enshrined in a map of the 135 different landscapes identified in Catalonia, that is, a map of 135 types of landscapes that are all essentially different.

1.1.1 Catalan Landscape Maps

The drawing up of the seven landscape catalogues has generated the first ever map of the 135 landscapes (known technically as 'landscape units') found in Catalonia. Catalonia has a similar surface area to states such as Belgium, the Netherlands and Taiwan, and possesses an extraordinary variety of landscapes that have developed as a result of its

[1]The Convention is applicable to the 46 member states of the European Council, that is, over 800 people in 26 countries that have to date ratified the agreement (including the Spanish State since 2008).

[2]Law 8/2005, 8 June, for the protection, management and organisation of the landscape (DOGC N0. 4407—16.6.2005). This execution of this law is guaranteed by Decree 343/2006, 19 September, for the development of Law 8/2005, 8 June, for the protection, management and organisation of the landscape, and regulates impact assessment reports and studies and landscape integration policies (DOGC 4723—21.9.2006).

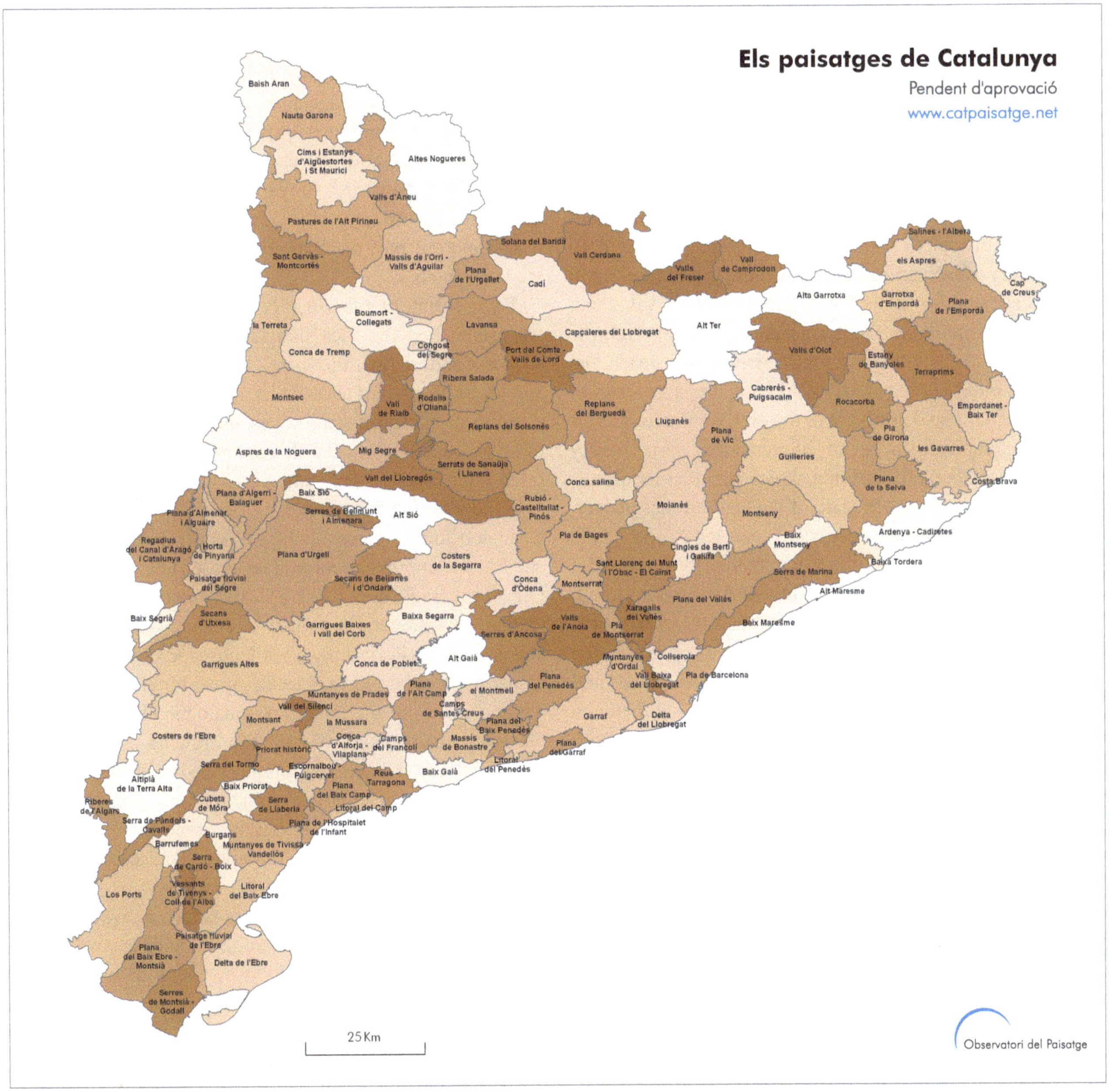

Fig. 1.1 Map of the landscapes of Catalonia. *Source* Catalan landscape observatory

particular relief features, climatic variation, natural ecosystems, historical legacy and cultural personality.

A bird's-eye view is enough to gain an immediate idea of the important contrasts that exist in the landscapes of Catalonia (Fig. 1.1). In the Pyrenees, the ecological and natural value of the small river valleys, the forests and the high pastures are evident, as are the extraordinary landscape structures present in the valley bottoms. As we approach the Lleida plains from the great bastion of the Pyrenees, the situation changes radically: here the ochres and yellows of the dry fields stretch seemingly beyond the horizon, a monotony only occasionally broken by small hills on which the local population has historically built its settlements. Southwards, the landscape changes again and becomes dominated by the limits imposed by and derived from the river Ebro and, for example, dry-stone walls and huts abound throughout the area. The Ebro Delta is a true icon in this landscape for both local inhabitants and all Catalans who appreciate not only its natural and ecological values but also the agricultural landscape that has been sculpted, of which the best-known feature today are the rice paddies. At the mouth of this great river its range of chromatic tones, its luminosity and horizontal scope, as well as a sensation of tranquillity, all warrant great respect. From the delta, the mountains that act as a frontier to this area of southern Catalonia are clearly visible, with the great massif of Els Ports, beckoning but still poorly known, in pride of place. Northeast, we head into the province Tarragona, in whose interior Mediterranean landscapes dominate and where—above all, in the county of Priorat—some of the most appreciated wines in the country are produced. Throughout practically the whole area, but particularly around Poblet and in the mountains of Vilobí and El Tallat, Les Garrigues Altes and El Montsant, dry-stone vernacular architecture is omnipresent in the landscape and furnishes it with great aesthetic value that is part of local heritage. Further north, passing through the Penedès region, also famous for its wines, we reach the Barcelona conurbation, which would require a chapter to itself. To the north and northeast of the city, we find two other landscape complexes of great interest: the agro-forestry mosaic of inland central Catalonia, which occupies thousands of hectares and is in a good state of health, and the greener, more humid and internally heterogeneous landscape of the counties of the province of Girona, where the landscape of La Garrotxa volcanic zone—amongst others—springs to mind. We could continue and increase the number of landscapes if we changed the scale of analysis. Initially, we would find the 135 aforementioned landscapes, perfectly described and delimited by the Catalan landscape catalogues, and then at a finer scale we would be faced with an extraordinary abundance of local micro-landscapes.

Each of these 135 landscapes is different from all the others. In other words, they are discrete areas, each characterised by a series of elements that bestows a certain degree of idiosyncrasy and sets them apart from all other areas of Catalonia, thereby ensuring that they are different from the other landscapes (but not any better or worse). These elements derive from an interaction of many factors such as climate, the combination of relief features (mountains, valleys, plains and so on), vegetation, land uses (arable land, built-up areas, riparian woodland, etc.), the organisation of the space, heritage, human activity stretching back centuries (building of roads and other more modern structures), perception (textures, colours, shapes, visibility), and existing landscape dynamics. The nature of the landscape is also determined by the relationships that are established between local people and their landscapes, which are unlikely to be revealed by a simple juxtaposition of maps (proximity, emotional ties, feelings of belonging, etc.).

In order to bring them closer to local citizens, these landscape units were given names based on the most popular local usages thrown up by the public consultation and participation processes that took place during the definition of the landscape catalogues. Thus, all 135 landscapes were named according to well-established local traditions that form part of local collective memories; we avoided technical or academic terms that are too divorced from local social realities. Each landscape has its own personal name, simply because the cities and their outskirts, the agroforestal mosaics, the vegetation, the architecture, local place names, smells and sounds, the sense of touch and other sensorial impressions all configure units that are peculiar to one particular place and nowhere else. For example, names such as Valls d'Olot, Moianès, Pinyana, Vall Baixa del Llobregat, Conca de Tremp and Alta Garrotxa are found in common usage and are part of local collective memory; that is why we used them.

By ensuring that the map covered the whole of Catalonia we established the idea that in landscape policy all landscapes—and not just the 'best', but also the most singular—should be taken into account. A map of this type enables us to turn our backs on stereotypes and to highlight landscapes of great value and interest that are generally little known and often not properly appreciated.

Landscape units are relevant because they constitute the basic territorial 'pieces' where specific landscape policies need to be applied according to the characteristics of each one. On average, they cover 237 km^2 and can be integrated—in fact, this process is already occurring—into both partial and master territorial plans via landscape directives, that is, landscape regulations that are designed for each territory and which stem from the landscape catalogue. This map is also becoming increasingly useful for drawing up or revising municipal urban plans. It puts knowledge at the service of action and has become an obligatory reference source in tourism, cultural and communication projects. It is also being used locally as a resource for promoting landscape projects of all kinds.

The landscape map of Catalonia is also a powerful pedagogical tool that will enable young Catalans and Catalan society in general to become more aware of the landscapes in which they live, and of the importance of their natural, cultural, social, productive, symbolic and identitary values, their singularity and the factors that are putting them at risk. Thus, the landscape map is now being used in some high-school textbooks as a topic in social sciences. It has also been used as the basis for the creation of educational materials for the middle-school topic *City, Territory, Landscape*.

In the landscape catalogues, each of the 135 types is accompanied by a complete datasheet[3] that describes its main characteristics, historical evolution, dynamics and the artistic expression that they have stimulated, as well as the threats that are currently hanging over it. The only other countries to have created similar landscape maps are Great Britain, France and Slovenia.

[3]The map can be downloaded from the website of the Catalan Landscape Observatory (www.catpaisatge.net).

1.2 The Character of the Volcanic Landscape of La Garrotxa

La Garrotxa volcanic zone lies chiefly in the landscape unit called 'Valls d'Olot' (Fig. 1.2), as defined in the *Landscape Catalogue of the Counties of the Province Girona*; this unit is centred on the county of La Garrotxa, but also embraces a small part of the counties of Pla de l'Estany and El Ripollès. The volcanic zone also contains two fragments of the landscape units of Rocacorba and La Plana de la Selva, as explained below.

With a surface area of 40,935 ha, the landscape of the Valls d'Olot encompasses all or part of the municipalities of Argelaguer, Besalú, Castellfollit de la Roca, La Vall de Bianya, La Vall d'en Bas, Les Planes d'Hostoles, Les Preses, Mieres, Montagut i Oix, Olot, Riudaura, Sales de Llierca, Sant Aniol de Finestres, Sant Feliu de Pallerols, Sant Ferriol, Sant Jaume de Llierca, Sant Joan les Fonts, Sant Miquel de Campmajor, Santa Pau, Serinyà and Vallfogona del Ripollès (Fig. 1.3).

1.2.1 Organisation and Current Landscape Dynamics

The landscape of the Valls d'Olot is the product of the configuration of a singularly complex territory in which tectonics and vulcanism have both played notable roles (Fig. 1.4). The mountain ridges and massifs such as the Serra de Malforat (915 m) that close the valleys off to the north belong to the first pre-Pyrenean ranges, whilst the ridges to the south of the region are part of the Sistema Transversal. Of this latter mountainous system, the main relief features in the Valls d'Olot are the east-west-running ridges of Serra del Corb (934 m) and Serra de Finestres (1026 m). These mountain chains act as a barrier between

Generalitat de Catalunya
Departament de Política Territorial i Obres Públiques
Observatori del Paisatge

Catàlog de paisatge de les Comarques Gironines
Unitat de paisatge 23: Valls d'Olot

Figura 23.24: Cinglera basàltica de Castellfollit de la Roca.

d'Olot. Al fons del carrer, trencant l'horitzontalitat, despunta el campanar amb coberta piramidal de l'església de Santa Maria. Destaquen també els fons escènics del Puigscalm, Santa Magdalena i els Plans de Falgars.

La dimensió estètica és innegable a raconades com els paratges de la Moixina, un dels més humits de tota l'àrea. Aquí, una part de la colada de lava del Croscat que arribà fins el Fluvià es va enfonsar en els flonjos sediments d'aquesta zona, donant com a resultat un paisatge on les roques volcàniques emergeixen, en alguns indrets, per damunt dels sediments i de l'aigua, que és molt abundosa a tota la raconada. Les fonts són nombroses i els recs de drenatge i els aiguamoixos s'alternen arreu; no falten els murets de pedra seca per salvar desnivells.

La intervenció humana sobre el paisatge ha estat igualment reeixida en altres indrets de la ciutat d'Olot com la plaça Clarà, l'eixample Malagrida, el passeig de Barcelona o el parc Nou. En aquest últim, de poc més de 3 ha. de superfície, simbiosi entre parc urbà, jardí botànic i espai d'alt valor ambiental i paisatgístic per la diversitat d'espècies vegetals que aplega i que convida al visitant a un passeig per deleitar-ne els sentits.

En els efectes estètics, a nivell cromàtic, intervenen les gradacions estacionals, especialment pel que respecta al cicle dels vegetals. Aquests efectes són ben palpables en paratges com la fageda d'en Jordà, que passa d'unes tonalitats verdoses ben vives i intenses durant la primavera i l'estiu, a uns tons ocres i marronosos durant la tardor, per perdre totalment les seves fulles durant l'hivern.

Als entorns fluvials els efectes estètics es multipliquen pel joc que s'estableix entre les làmines d'aigua i una vegetació esponerosa com és la de ribera. A part dels salts naturals i de les gorges, l'existència de velles rescloses molineres ha redundat en la creació d'algunes d'aquestes làmines, en ocasions, aprofitades pel bany. Es poden citar el salt dels Portuguesos, el salt del Sallent a la vall d'en Bas o la bassa del molí d'en Murris a les proximitats de Sant Feliu de Pallerols, les gorgues del pla d'en Xurri, al riu Gurn, aigües amunt de Sant Privat d'en Bas o la zona del Tussols - Can Basil al terme municipal d'Olot. Els salts d'aigua són especialment valorats ja que aporten dinamisme al paisatge. Cal també destacar l'interès estètic del tram del Fluvià que transcorre entre Les Preses i el pont de les Mòres al municipi d'Olot.

Figura 23.25: El Puigsacalm.

L'ocupació i l'activitat humanes que al llarg dels segles s'han produït sobre un territori han conferit al paisatge uns **valors històrics**. Hi ha un patrimoni construït que utilitza la matèria primera local i que arriba a perfilar elements identificables del paisatge. La roca volcànica s'ha emprat per a l'edificació de cases i masies, per cabanes de pastors i murs de pedra seca, és a dir, sense la utilització d'argamasses. Les principals àrees amb construccions de pedra seca comprenen els voltants d'Olot, sobretot l'àmbit del municipi de les Preses, seguint les colades volcàniques. Es caracteritza pels seus murs i clopers fets amb pedra d'origen volcànic, que delimiten antics camps agrícoles de petites parcel·les. També s'hi troben cabanes, algunes d'elles incorporades als mateixos murs de pedra seca. Els dos sectors de les Valls d'Olot en els quals la construcció en pedra seca gaudeix d'uns millors acabats visuals són tant a l'altiplà basàltic de Batet com a sobre la colada de lava del bosc de Tosca o entre Olot i Les Preses. Les dimensions, colors, formes i textures dels fragments de les roques basàltiques varien d'acord amb la seva gènesi. El 1764 s'autoritza el conreu en el bosc de Tosca. Fins llavors era un gran erm ple de roques basàltiques, escòries volcàniques i laves. Les pedres i el pedruscall s'amuntegaren per permetre el conreu de la terra i, acte seguit, es formaren els murs per delimitar les parcel·les i regularitzar els marges de fort desnivell. També s'aixecaren barraques per guardar eines, com a sopluig pels pagesos o pels animals.

Més enllà d'aquestes construccions més modestes, es troben conjunts arquitectònics edificats amb roca volcànica que tenen un valor monumental elevat i que són punts de referència en el paisatge de la zona. S'ha de citar aquí la Vila Vella de Santa Pau, declarada conjunt historicoartístic l'any 1971. La Vila Vella ofereix unes característiques medievals molt marcades i s'estén per un turó proper al riu Ser. Queda delimitada per l'antiga muralla dins la qual es distribueix un conjunt de cases que semblen acomodar-se al castell datat del segle XIII. Destaca també l'església de Santa Maria, del segle XV, amb el robust campanar quadrat i tot el grup de carrerons estrets, amb angles sobtats i descensos pronunciats, que desemboquen al Portal del Mar. Per a controlar la relació entre aquest conjunt i les zones urbanes i els espais paisatgístics que li donen suport, la Generalitat de Catalunya va establir, l'any 2000, un entorn de protecció que inclou tot el turó.

Hostalets d'en Bas, al cor de la plana de la vall d'en Bas, també està declarat com a conjunt històric-artístic. El seu interès recau tant en la tipologia unitària de les cases com en els materials de construcció. D'altra banda, també al bell mig de la vall, però en un turó de poc més de 70 metres d'altitud se situa la capital històrica de la vall, El Mallol, en la que encara es pot admirar un esplèndid conjunt arquitectònic tot i haver perdut la seva fortalesa.

Altres conjunts arquitectònics que no poden ser obviats són el de Sant Privat d'en Bas, nucli medieval al qual s'accedeix per un passatge porticat; Joanetes, aturonada als vessants de la vall d'en Bas i amb l'altívol perfil del Puigsacalm com a teló de fons; Sant Feliu de Pallerols,

Figura 23.26: Cràter i ermita del volcà de Santa Margarida.

471

Universitat de Girona
Universitat Politècnica de Catalunya

Fig. 1.2 Example of the datasheet of the landscape unit 'Valls d'Olot' included in the *Catalogue of the Landscapes of the Counties of Girona*. *Source* Catalan Landscape Observatory

Fig. 1.3 Map of the landscape units present in the counties of Girona, with the Valls d'Olot, Rocacorba and the Plana de la Selva highlighted. *Source* Catalan Landscape Observatory

the land that drains northwards into the river Fluvià, and that that drains southwards towards the river Ter. To the east rises Serra de Sant Julià del Mont (907 m), which, together with its secondary spurs such as Serra Cugulera and Serra del Torn, constitutes a veritable orographic wedge standing between the Santa Pau and Mieres basins, on the one hand, and between Santa Pau and the Fluvià valley, on the other. The river Fluvià and its tributary, the Ser, drain this sector of this landscape unit.

Tectonics have played a key role in the modelling of the current relief of the Valls d'Olot. The dissecting of the territory by a series of perpendicular faults has given rise to a horst and graben system, that is, alternating raised blocks (horsts) and sunken valleys (grabens). The former correspond to some of the main mountain ridges in the area, while the latter are represented by the most important valleys and basins. The two principal fault lines run north-south, one located at the foot of the mountain of Puigsacalm and the other between the mountains of La Garrotxa and the Empordà plain. Between these two faults, a system of perpendicular secondary faults has developed that delimits the raised blocks of Rocacorba, Serra del Corb, Serra de Finestres and Serra de Sant Julià. These faults have also provoked the appearance of sunken blocks that are occupied by small valleys, troughs and basins. The main such feature is the Olot-Mieres Trough, which has been subdivided by the action of small faults running parallel to the main fault into three basins and secondary horsts: from west to east, these are the Olot Basin (420 m), Santa Pau Basin (450 m) and Mieres Basin (250 m), separated respectively by the small raised blocks of Santa Llúcia (670 m), Montestí (688 m) and Montfalgó (614 m).

Fig. 1.4 Panoramic views of the Valls d'Olot. *Source* Catalan Landscape Observatory

Tectonic movements and the system of faults that they have originated have given rise to a number of eruptive magmatic crises over the past 350,000 years. This volcanic activity has generated a series of relief forms that modified pre-existing relief features and hydrological networks and now constitute the best-preserved volcanic region in the whole of the Iberian Peninsula. The most recent volcanic episode—the eruption of the volcano Croscat—took place 11,500 years ago. The volcanic activity in this area has enriched the variety of the area's relief with a series of 40 volcanic cones, mostly Strombolian in nature, that range from 10 to 160 m in height and 300–500 m in diameter. A total of 40 cones and 20 lava flows are the most evident geological and landscape features that bear witness to this past volcanic activity. The features that stand out most in the landscape are the volcanic cones, which rise up in valley bottoms or in more or less flat areas (e.g. the volcanoes of La Garrinada, El Montsacopa, Croscat and Santa Margarida), and the basaltic lava flow on top of which the village of Castellfollit de Roca was constructed. Also part of this volcanic zone are the cones that are less visible due to being encrusted onto mountainsides or ridge tops; in addition, a small number of less spectacular volcanoes are found in the valleys of the rivers Brugent and Llémena.

The different types of flows and cooling processes that took place in the lava flows have given rise to curious forms in the basalt cliffs, which have been cleaved into columns, blocks and slabs depending on whether the lava cooled whilst still in motion or once it had stopped. The best known such features are found in the lava flows at Castellfollit de la Roca and Sant Joan les Fonts. In all, volcanic materials cover over 3300 ha of the area. The lava flows emitted by the volcanoes acted as a modelling force on the relief and landscape forms: they flowed down slopes and occupied river valleys, filling them with volcanic material and converting them into fertile plains. The most significant example is that of the formation of the Vall d'en Bas, billiard-table flat due to the natural dam formed at the entrance to the valley by the basaltic lava flow emitted by the volcano of El Racó. This barrage modified the drainage of the valley and created a marshland behind it that eventually silted up. Some marshes still exist in badly drained areas such as the Font Moixina and Can Jordà.

The landscape of the Valls d'Olot has also been affected by relief features with slopes that occur on ridges constituted of alternating layers of sedimentary materials (marls, mudstones, conglomerates) that have differing resistance to erosion. These reliefs have steep cliffs on their scarps but much gentler dip slopes that have formed on surface strata that are more resistant to erosion. This type of morphology is present in Serra de Sant Julià del Mont, Serra del Corb and Serra de Finestres.

The combination between the fractured relief and the hydrological network that drains the area has defined a series of valleys of various size that principally drain into the river Fluvià. Indeed, a large part of this river's upper course and many of its tributaries flow within the Valls d'Olot landscape unit. The role of this river and its tributaries in the structuring of the landscape is accentuated by the pattern of this hydrological network. Initially, the Fluvià runs south-north but then turns eastwards decidedly and runs west-east to the Mediterranean. Thus, over half of the surface area of the Valls d'Olot —above all in its western and northern sectors—is centred on this great depression drained by the river Fluvià and its tributaries, and recognisably discrete sectors are known, for example, as the Vall d'en Bas or Vall de Bianya, or constitute the valley of the mid-course of the river Fluvià itself. Even despite the whims of geology, it is still possible to discern clear physical similarities between these three valleys. In the most eastern part of the area, the valley of the river Ser, flows into the Fluvià outside the area defined as the Valls d'Olot.

The situation of the Valls d'Olot in the heart of the humid backbone of mountains that extend from the Vallespir in the Pyrenees to El Montseny guarantees higher rainfall and a less marked summer drought than in neighbouring regions. Thus, in the valleys at 400–500 m a.s.l average annual rainfall is over 1000 l/m^2, a figure that in Catalonia is normally only recorded in places over 1000 m a.s.l. The mountain of Puigsacalm (1515 m a.s.l.) rises perpendicular to the moist prevailing winds that blow off the Mediterranean and acts as a barrier that ensures increased rainfall in the western sector of the Valls d'Olot. Furthermore, the great barrier of the Pyrenees to the north acts as a shield that protects the area from the cold northerly airstreams.

These climatic conditions help explain the vegetation of the Valls d'Olot. Firstly, the thermal inversions that occur in the troughs—caused by the presence of cold air in the valley bottoms—give rise to regular fogs. Along with the high rainfall, this phenomenon explains why many plant species such as beech (*Fagus sylvatica*), birch (*Betula pendula*) and pedunculate (*Quercus robur*) and sessile (*Quercus petraea*) oaks that are more typical of moist central European climes thrive in the landscapes of La Garrotxa. Nevertheless, Mediterranean environments are far from absent in the area. The Valls d'Olot boasts a variety of plant communities whose presence or otherwise is determined by altitude and, above all, by aspect (contrast between north- and south-facing slopes); a further factor to take into account is the gradation from the easterly parts of the county, more influenced by Mediterranean environments, to the more humid western sector where precipitation is greater. Thus, in the easterly part of this landscape unit in and around the Mieres basin the dominant forest type is holm oak with laurustinus (*Viburnus tinus*), wherever it has not been substituted by Aleppo pine (*Pinus halepensis*) woodland.

In more elevated areas such as the Serra de Sant Julià del Mont and Serra de Finestres, continental holm oak forests extend across mountainsides, while further west in more humid areas such as the Serra de Marboleny, Serra del Corb, Sant Miquel del Mont or the western part of the Vall de Bianya, oak forests come into their own and beech forests clothe the highest peaks. An exception to this rule is the famous D'en Jordà beech forest, which covers a transitional area between the Santa Pau and Olot basins at an altitude of 500–600 m a.s.l. Sub-Mediterranean oakwoods of downy oak (*Quercus pubescens*) cover large areas, while in and around Olot humid oak woodland and mixed forests of pedunculate oak, large-leaved ash (*Fraxinus excelsior*), field maple (*Acer campestre*) and limes (*Tilia* sps.) grow on the basalt lava flows. Overall, these forests—above all, those found on mountainsides and most of those within the boundaries of La Garrotxa Volcanic Zone Natural Park—tend to form large continuous masses of dense forest.

In valley bottoms and flatter areas, the natural vegetation has been substituted by fields, pastures and built-up areas. Agricultural areas such as the Mieres and Santa Pau basins (Fig. 1.5), the Vall de Bianya and the sedimentary terraces in the Fluvià valley downstream from Castellfollit de la Roca are largely devoted to growing fodder crops for cattle. However, urban growth and the expansion of the road network is tending to fragment this arable surface and create patches of land that are often abandoned and left as wasteground. Despite factors such as the great fertility of most of the county's soils and the relatively accident-free relief that favour the survival of local agriculture, potential agricultural soil is being lost in the urban area around Olot and alongside the main infrastructures. On a smaller scale, the Pla de Sant Miquel del Corb and the Brugent, Cogolls and Colltort valleys boast good arable land, perfectly adapted to the terrain, that is not threatened by the dynamics of urban growth.

Agriculture in the Vall d'en Bas is in vigorous health due to the perfect conditions for the cultivation of cereal and fodder crops that give high yields. Nevertheless, this agrarian landscape is becoming increasingly homogenous due to agricultural intensification and specialisation in animal husbandry, as well as the promotion of crops such as maize. Potatoes are making something of a comeback in the Vall d'en Bas and leguminous crops—above all, haricot beans—are also still important.

Extensive animal husbandry continues to exist in the mid-to-high mountains. By contrast, in the valley bottoms cattle are generally kept indoors in stalls, easily identified by their modern designs and often built right next to centuries-old farmhouses from pre-fabricated materials.

Smaller, more undulating open areas such as the basaltic plateau of Batet de la Serra retain the best examples of

Fig. 1.5 Agroforestal landscape around Santa Pau, consisting of small flat areas lying between the volcanoes. *Source* Catalan Landscape Observatory

balanced mosaic landscapes that consist of patches woodland, arable land and pastures linked by an interwoven network of paths lined, in many cases, by walls of volcanic stone that back onto the very farmhouse buildings. Another area that exhibits a combination of habitats are the volcanic cones, where thickly forested slopes look down on the bottoms of craters occupied by pastures and meadows. A few small vestigial open spaces in upland areas are still devoted to extensive cattle grazing. The abandoning of these more marginal areas inevitably leads to the spread of woodland and the formation of ever more continuous forest masses.

If not accompanied by corrective measures, the intensification of agropecuary practices could provoke a gradual loss of some of the features that characterise the rural landscape of the Vall d'en Bas. Some of the factors that will point to any such degradation include the proliferation of farm out-buildings and other constructions that do not match the historical local buildings, the disappearance of hedgerows and the contamination of the groundwater.

Along the terraces of the valley of river Fluvià and near many town centres (above all, where water can be channelled), family-run vegetable gardens of different shapes and sizes are still tended by local people (e.g. Castellfollit de la Roca, Sant Jaume de Llierca, Olot, Besalú and Les Planes d'Hostoles). Along the river Fluvià, above all between Olot and Castellfollit, some of the factories built at the end of the nineteenth and beginning of the twentieth centuries still stand, with intact associated systems of canals and weirs, mills and small HEP plants. Small plantations of trees extend along the banks of the river Brugent in a few places.

The flattest areas hold almost all the built-up land in this landscape unit. Today's urban landscapes reflect a territorial organisation dating back to the creation of rural parishes and medieval fortifications that, in cases such as Santa Pau, Besalú, El Mallol, Sant Privat d'en Bas, Sant Feliu de Pallerols, Les Planes d'Hostoles and Riudaura, have evolved into important historical urban nuclei. Many of this unit's smallest settlements have changed little and the buildings of these farmsteads and hamlets remain clustered around their churches. Larger villages have streets of a certain length, to which rows of modern housing have been added; this has created disorderly urban nuclei that enclose and isolate non-built-up agricultural spaces, as in the case of Sant Feliu de Pallerols and Les Preses. Such spaces are more frequent in flat areas than in old village centres (e.g. Sant Esteve d'en Bas) or alongside new communication infrastructures (e.g. Les Preses, Les Planes d'Hostoles, Castellfollit de la Roca and Santa Pau).

A rather different case is that of the city of Olot, whose urban landscape is the sum of the successive growth phases that have taken place over the centuries. It has expanded in all directions, but above all into areas where the relief is less of an obstacle and along the main streets and roads towards Castellfollit de la Roca, Amer, Riudaura and Ripoll. Today, the types of constructions on display here are highly varied and occupy the land in a chaotic fashion; built-up areas mix in with agricultural areas without any apparent structure or organisation. This sense of dispersion is intensified by the presence of out-of-town industrial estates whose great depots rise up next to the main roads and road junctions.

Despite the wealth of new housing in Olot, the towns in its hinterland are today witness to a growing process of 'suburbanisation'. From a perspective of the physical transformation of the space, the result is, firstly, greater occupation of the small open areas on the outskirts of Olot (Pla de Dalt, Pla de Baix) and the separation of intermediate agricultural areas into isolated plots surrounded by built-up areas. Secondly, aside from some areas that still maintain their predominantly open feel, an urban continuum with no room for continuation is beginning to extend over the eight kilometres that separate the town of Les Preses in the south from La Canya, Llocalou and Sant Joan les Fonts in the north. At the same time, the entrances and exits to the built-up area of Olot have arisen with little aesthetic coherence and are destined to act as mere shop-windows for commerce and services. This trend is obvious along the entrances to Olot via Les Tries and the roads from La Canya and Sant Joan de les Abadesses. The same phenomenon is occurring in Besalú.

The combination of industrial estates and large out-of-town commercial centres is most frequent along main roads and in the vicinity of built-up residential areas, which essentially correspond to the entrances into Olot; at the same time, these spaces are also witnessing the consolidation of a number of light industrial estates, for instance, in the Vall de Bianya and along the lower course of the river Fluvià (e.g. Pla de Politger in Sant Jaume de Llierca, Les Serres industrial estate in the Vall d'en Bas, and Besalú).

Restrictions on this model of territorial occupation are fixed by the boundaries of La Garrotxa Volcanic Zone Natural Park and, to a certain extent, by relief features such as the volcanic cones of La Garrinada, El Montsacopa and Montolivet that now stand out as small fragments of natural land surrounded by a much larger urban area. It is worth highlighting too the appearance in the Vall d'en Bas of residential complexes that, despite their small size, fail to blend in with pre-existing local buildings. The fact that they have been built in a geographical area in which the stamp of urban growth to date is all but inexistent underlines their visual impact.

The main roads connecting the Valls d'Olot with neighbouring towns and cities run along the valley bottoms. The Vall d'en Bas is crossed by the C-153 Vic-Olot trunk road that leads to the entrance to the Bracons tunnels; the Vall d'Hostoles is traversed by the C-63 to Santa Coloma de Farners; the C-26 runs through the Vall de Bianya and provides access to the county of El Ripollès; and the valley of the Fluvià downstream from Olot is home to the A-26 dual carriageway (the old N-260) to Besalú, which links to the roads to Figueres and to Girona via Banyoles.

The extending of this road network in recent years has led to a transformation of the landscape of the Valls d'Olot. The complex relief that encompasses these valleys meant that for many years the only good road link with the wider world that did not involve crossing various mountain passes or having to negotiate a series of tight bends was northeast along the valley of the Fluvià. However, from the 1990s onwards major reforms began to be made in the road network starting with the remodelling of the road through the Vall d'Hostoles and the Bas tunnel, subsequently followed by the construction of a by-pass on the N-260 north of Olot and the new C-153 through the tunnels of Collabós (or Capsacosta) that gives much better access to El Ripollès around Sant Joan de les Abadesses.

Nevertheless, the two road projects—which are closely connected—that have had most impact on the territory and landscape are the widening of the Besalú-Olot (A-26) road into a dual carriageway (with a future connection to Figueres) and, above all, the new trunk road between Vic and Olot passing through the Bracons Tunnel (C-37) (Fig. 1.6). The impact of the former road on the landscape is patent, basically because of its width (four lanes), its succession of embankments, cuttings and viaducts, and the need to build junctions with the pre-existing road network. It is has led to a notable loss in ecological connectivity and landscape quality between the areas of natural interest of L'Alta Garrotxa, on the one hand, and the mountain of Sant Julià del Mont and the La Garrotxa Volcanic Zone Natural Park, on the other. The impact of the C-37 is above all manifest in the Mas Rubió viaduct that crosses the river Fluvià and links the tunnel of La Codina to the agricultural plain of the Vall d'en Bas. In the immediate future, the exact route of this road through Les Preses and around Olot, as well as its connection to the A-26 dual carriageway, is still to be decided.

Of the power lines that cross this landscape unit, the largest is the 132-kV line that follows the river Llémena and crosses the Serra de Finestres to the east of Finestres Castle, before heading for Santa Pau and Olot and the sub-station situated between the volcanoes of La Garrinada and El Montsacopa. This power line probably has a greater impact on the landscape and the environment than any other such infrastructure since it cuts through the heart of La Garrotxa Volcanic Zone Natural Park from southeast to northwest.

It is worth emphasising the role La Garrotxa Volcanic Zone Natural Park has played since it was created in local landscape dynamics. The permanence and effectiveness of this vital institution in the management of local natural systems is key in the maintenance of the area's landscape and environmental quality, and in ensuring that local economic activities, above all agriculture and tourism, are compatible with the protection of the environment.

1.2.2 Historical Evolution of the Landscape

In comparison with other nearby geographical areas such as the Pla de l'Estany, the first indications of any prehistoric culture in the Valls d'Olot are relatively late. Remains classified as Palaeolithic have been recovered from a site at Pla de Politger (Montagut). Much clearer are the clues relating to Iberian and, above all, Roman settlements; the latter civilisation left an obvious mark on the landscape in

Fig. 1.6 The entrance into Olot from the south. *Source* Catalan Landscape Observatory

the form of agricultural techniques and the first communication infrastructures. In the latter case, the presence of the Roman Via Annia, which followed the course of the river Fluvià, can still be traced at the head of the Vall de Bianya. This road was used for many centuries after the Roman occupation had finished and throughout the Middle Ages enabled settlers to penetrate inland.

Feudalism heavily influenced the new styles of human settlement and had a serious impact on ecclesiastical and noble prerogatives. Monasteries such as Santa Maria de Riudaura and Els Arcs de Santa Pau were founded to help promote the repopulation of the territory. Besalú, the capital of the eponymous countdom, grew up around the castle and church of Santa Maria, to which buildings such as the monastery of Sant Pere and the notable medieval bridge over the river Fluvià were added at a later date.

Numerous small isolated churches, many of which still stand, were consecrated and acted as the focal point for numerous fresh settlements that, in many cases, have survived up to the present day. Likewise, many castles such as those of Finestres, Colltort, Castellfollit, Bas or Castelló, El Mallol and Santa Pau—some of which are still standing—were built to protect the frontiers of the local feudal domains.

The essential ways of life that have survived up to the modern period were defined in the Middle Ages. People's lives and associated tasks were structured around the farmsteads, whose influence on the landscape was double: from an economic perspective these farms fomented the exploitation of natural resources, whilst architecturally they were the most representative element of rural building styles. Around Olot, the use of volcanic materials as building stones was commonplace; in the Valls d'Olot in general the oldest farms are situated on the lowest slopes of the mountains, preferably on the sunnier south-facing slopes, where they were perfectly placed to exploit the nearby forests for produce and harvest the crops growing in the agricultural plains. The density of farms falls with altitude but as the land became depopulated the farms at greater altitude were abandoned first.

From the Middle Ages onwards the ploughing up of the land to plant crops in both valley bottoms and on small patches of flat land in upland areas began to gain momentum. Between the ninth and eighteenth centuries crops—albeit with a certain degree of diversity—were generally uniform in type. Autumn cereals (e.g. wheat) were the most widespread of all crops, followed by oats and barley. A spring crop, sown primarily on the more humid volcanic soils in the west of the region, was buckwheat, whilst on drier, non-volcanic soils in the east millet was common. In the ninth to fourteenth centuries, due largely to local subsistence production on farmsteads, vines covered large areas of the territory, above all in eastern areas with more Mediterranean climates. From the nineteenth century onwards, olive cultivation expanded notably in the drier eastern sector and there were also large areas of pastureland.

However, the most significant transformation of the agricultural landscape took place at the end of the eighteenth century with the introduction of maize and, to a lesser extent, potatoes into the area as crops. Maize cultivation meant that fields were no longer allowed to lie fallow; at the same time olive and, above all, vines began to enter progressively into decline. Wood-charcoal production peaked in the nineteenth century; forests were also exploited, albeit to a lesser extent, for firewood and timber, the latter used, amongst other things, for building beams. The intensification of forestry exploitation, spurred on by local industrialisation, severely damaged local forests, which shrank alarmingly and progressively, as is graphically illustrated by many of the photographs taken at the end of the nineteenth or beginning of the twentieth centuries.

Demographic growth in the eighteenth century (and to a lesser extent the nineteenth century) obliged farms to grow and new buildings to be constructed, with which the density of the local scattered human population also grew. For the first time, large-scale agricultural production got underway in the Vall d'en Bas and in the nearby Bosc de Tosca (Fig. 1.7). The towns in the valley bottoms expanded, a process that in Olot sparked the construction of new urban areas outside the old city walls. Besalú, on the other hand, lost both demographic and economic sway to Olot. Industrial growth of textile factories, paper mills, modern flour mills and HEP stations changed the nature of certain fluvial landscapes. This occurred above all along the river Fluvià between Olot and Sant Jaume de Llierca, where the narrower stretches of faster-flowing water enabled systems of weirs, canals and ponds to be built to extract the maximum energy from the river's waters.

Fig. 1.7 The Bosc de Pedra Tosca. *Source* Catalan Landscape Observatory

In the second half of the nineteenth and the first quarter of the twentieth centuries, the Valls d'Olot began to enjoy improvements in its communication network that finally condemned the traditional network of mule tracks and 'royal' roads to the past. Thus, in 1849 the first road linking Olot to Girona via Besalú was opened, followed by the Olot-Sant Joan de les Abadesses road in 1880, and the Olot-Vic road in 1925. In the twentieth century, the roads between Olot and the neighbouring towns and villages were built, along with the direct road to Ripoll and, amongst others, the road from Camprodon to Girona. Many stretches of these roads in fact left a harmonious imprint on the landscape, even though most of these once bucolic tree-lined roads have now disappeared. Olot became the county's main communication node, a position reinforced by the narrow-gauge railway line (today a greenway for cyclists) that linked the city to Girona in the period 1911–1969.

The landscapes of the Valls d'Olot underwent serious changes in the second half of the twentieth century. Firstly, a sudden abandoning of forest exploitation took place due largely to the fall in wood-charcoal production provoked by the rising popularity of alternative fuel sources such as butane gas. Secondly, croplands and the farmsteads situated in the highest, least accessible areas were abandoned—only in the best of situations were these lands then devoted to seasonal grazing. By contrast, valley-bottom agriculture intensified and became more mechanised, the classic case being that of the Vall d'en Bas whose lands were consolidated at the end of the 1960s. Crop rotation was simplified and maize became by far the most important crop in the valley, aided by sprinkler irrigation. Typical crops such as buckwheat—which was of great relevance to the landscape as its fields were repeatedly depicted by the artists of the Olot School of Landscape Art—all but disappeared.

Olot and, to a lesser extent, other towns grew notably in this period and often occupied land extending beyond the early-twentieth-century new towns such as the Olot Malagrida garden city; by the 1950s whole new residential quarters separate from the town centres (e.g. Sant Pere Màrtir, Bonavista, Hostal del Sol and Morrot) were being built to house new residents. Industry also expanded and diversified, and stopped being so closely linked to the river valleys, while urban growth in general was mostly carried out at the cost of agricultural land. From the 1960s onwards, the population of Besalú also grew, as did its industry.

The intensification of the processes described above had a negative impact on the visual transformation of the environment. One of the most dramatic examples was the quarrying of volcanic material—the so-called *greda* or lapilli —to manufacture concrete and cement for construction. During the 1970s and 1980s, the morphology of a number of volcanic cones was seriously altered, an impact that increased local social awareness- of the on-going degradation of the volcanic zone. As a result, in 1982, the Catalan government declared the volcanic zone a Natural Site of National Interest and then, in 1985, a Natural Park. Today, the landscape and the discovery of its natural values have encouraged the creation of a series of facilities and services designed to welcome visiting tourists.

1.2.3 Artistic Expression in the Landscape

One place in which the landscape has undoubtedly influenced many types of artistic expression is the Valls d'Olot. Olot and its surrounding area have inspired numerous local painters, as well as many from the rest of Catalonia. The best known such artists were working during the Catalan artistic Renaissance—the 'Renaixença'—in the second half of the nineteenth century, when landscape painting came to be regarded as an avant-garde form of pictorial art. The Olot School grew in this period and was constituted formally in 1873 as part of the School of Fine Art (a public school of drawing), a branch of the School of Fine Arts in Barcelona. It was founded to train specialist artists to work on the design of the *indiennes*, printed textiles that required artists and draftsmen, and its existence ensured that Olot remained at the forefront of Catalan art for a number of years. Members of this school include eminent and influential painters such as the Vayreda brothers—Joaquim and Marià—and, above all, the painter Josep Berga i Boix, who became director of the school in 1877. Joaquim Vayreda, considered the father of the Olot School of Landscape Art, studied in Olot and then in Barcelona, and developed a style of landscape painting that was both direct and natural. Berga i Boix, a good friend of Vayreda's, was director of the Olot Artistic Centre, created in 1869, an institution that was a hotbed of artistic learning that remained somewhat apart from the official School of Art. Many painters from Olot and the rest of Catalonia—Caba, Urgell, Armet, Galofré, Pellicer, and, later on, Tamburini, Urgellès and Joan Llimona—were part of the Centre and became the nucleus of what has come to be known as the Olot School. From 1877 to 1914, Berga i Boix was director of the Public School of Drawing, whose name changed to the Technical School of Fine Art in 1891.

The Olot School principally painted landscapes, although some of its members did also produce other types of compositions and figures. The School identified with certain features of the landscape—rural, humanized and idyllic—and its output was designed to reinforce its ideological and symbolic beliefs. The most-depicted landscapes were the water meadows of La Moixina (Fig. 1.8), the Vall d'en Bas and the omnipresent mountain of Santa Magdalena. Due to their beauty, the county's riparian woodland was another favourite subject and by the beginning of the twentieth

Fig. 1.8 La Moixina by Marian Vayreda, one of the landscapes most painted by the Olot School. *Source* La Garrotxa County Museum

century such scenes were sufficiently well known to be used on postcards as souvenirs, sold largely to the many people who spent the summer in the area. The landscape of the Valls d'Olot was perceived by these artists as a group of shapes, masses and colours, whose totality they expressed in their pictorial compositions. The landscape artists of the nineteenth and twentieth centuries shared the need for direct, almost existential contact with the landscape and so often painted in the open air. When necessary, they exaggerated or even misrepresented certain aspects of reality to ensure that their works conformed with their concepts of morality.

Alongside the painting, the landscape of the Valls d'Olot also served as a source of inspiration for a number of authors. The volcanic landscape of craters and basaltic lava flows has inspired a thousand myths and legends, as well as supernatural stories that were transmitted orally through the generations until they are finally collated and transcribed.

At the beginning of the twentieth century, a number of writers used the Valls d'Olot to express their moral values. The paradigm of this idea was the mountain landscape, which, in the eyes of local writers, was a pure, virgin, sacred and intact repository of both moral and national values that endowed the region with its character and identity. In his *Illustrated Guide to Olot and its valleys. The Catalan little Switzerland* (1908), Mossèn Gelabert, a priest, wrote:

> [...] a bountiful spring, a basalt grotto... the course of the river Fluvià enclosed between cliffs of lava..., all this and many other things transport you unwittingly to a state of moral rectitude and joy, from which only a violent shock can wake you [...] Both the landscape and the vegetation that clothe it offer the walker uncountable rich and abundant reasons for study and for poetic and scientific inspiration [...] (Gelabert 1908).

The poet Joan Maragall provided a pantheistic and contemplative vision of nature. He sought to find where the human spirit coincided with that of the natural world and tried to find transcendence by immersing himself fully in nature. This is the sensation that he had in the heart of the D'en Jordà beech forest:

> Wayfarers as they enter
> Begin to walk more slowly
> Measuring their steps in the vast silence
> They stop, hear nothing, they are lost
> A soft forgetfulness of the world overcomes them
> In the silence of that deep place
> They think not of leaving or if they do it is in vain:
> They are captured by the Fageda d'en Jordà
> Prisoners of its silence and viridity
> What companionship! Oh liberating prison!
>
> (Joan Maragall 1911)

Many other authors such as Josep Pla have dedicated literary texts to the Valls d'Olot and its landscapes. The city of Olot itself, the river Fluvià and its tributaries, the many natural springs, and the natural surroundings of Olot are some of the most commonly described landscapes. According to Josep Pla, one of the best of all experiences is "to look out of a window and contemplate the surrounding landscape" of Olot. He wrote of the "evergreen fields, trees [...] dressed with their salt-and-pepper colours in autumn; an extremely agreeable tarnished ruddy background; and the black, brilliant hills with playful irregular outlines".

Olot's poetry is closely linked to the white flowers of the buckwheat. Many poems refer to this crop's flowers, which were also a source of inspiration for many painters. In 1908, the flower of the buckwheat was chosen as the symbol of the poetic encounters that were organised in La Moixina.

The landscape of the Vall d'Olot has been used in films such as *The perfume: The story of a murderer* (2006) by Tom Tykwer, of which some scenes were filmed in Besalú; *The knight of the dragon* (1985) by Fernando Colomo, filmed in the volcano Croscat; *Serrallonga: the legend of a bandit* (2008) by Esteve Rovira, which was partly filmed in Santa Pau; and, most recently, *The artist and the model* (2012) by Fernando Trueba.

1.2.4 Values of the Landscape

The following section describes seven types of values present in the landscape (ecological, aesthetic, historical, productive, social, religious and spiritual, and symbolic and identitary). Natural and ecological values refer to factors or elements that determine the quality of the natural environment; aesthetic values are related to the capacity of a landscape to transmit a certain sense of beauty; historical values consist of the most relevant of the traces that human activity has left on the landscape in the past; productive values are linked to the capacity of a landscape to provide economic benefits via the transformation of its constituent parts into resources; social values are associated with the uses that people or groups of people make of the landscape; religious and spiritual values correspond to elements of the landscape that, together, provide inspiration for religious practices and beliefs; and, finally, symbolic and identitary values are those that help a particular group of people empathize with a particular landscape.

In terms of natural and ecological values, the most essential element of this landscape unit are the volcanoes, which provide a unique legacy that is quite unlike any other anywhere in the Iberian Peninsula. This encouraged the Catalan government to pass a law in 1982 to protect La Garrotxa Volcanic Zone and then declare the area a Natural Park in 1985. Over the years, the Park has been modified and currently has a surface area of 15,309.4 ha. It is also home to 28 natural reserves (1180.4 ha), which receive special protection due to their inherent geological and botanical value and largely correspond to the most interesting—from a geomorphological perspective—volcanic cones such as those of Santa Margarida, Croscat and Roca Negra. The sum of the geological, biogeographical and climatic factors guarantee that the Park's flora is extraordinarily diverse: Mediterranean, sub-Mediterranean and central European elements combine to give a current total of 1173 vascular plants identified in the Park. Forests cover 65 % of the Park and are dominated by holm oak, deciduous oak and beech forests, although there are also a number of mixed forests and alder groves. The Park's management of these forests guarantees the preservation of this vital element in the landscape.

It is also worth highlighting the highly varied ecological importance of the river Fluvià, whose upper course runs through the Valls d'Olot. On the one hand, it is the only large river in Catalonia that is not damned anywhere along its course, a feature that bestows upon it a crucial role as an ecological corridor; on the other hand, this function as a corridor goes beyond the limits of the Valls d'Olot since this river also flows through the northern sector of the Pla de l'Estany and the Empordà plain, thereby linking the inland forests and upland areas in the Valls d'Olot with the coastal marshes of the Aiguamolls de l'Empordà. The river also helps reduce the barrier effect of some of the many road infrastructures that are present in the region.

The ecological values of smaller rivers such as the Riera de Bianya and Riera de Riudaura and their tributaries should not be underestimated. These watercourses run west-east and, after crossing a series of often large agricultural plains, link the forests of their headwaters in the mountains of Santa Magdalena and Capsacosta to the Olot basin and the river Fluvià. Near its source, the Riera de Bianya in fact is divided into three tributaries: Riera de Sant Ponç, Riera de Farró and Riera de Santa Llúcia de Puigmal, a fact that increases the impact that the main river has on the surrounding landscape.

On a smaller but complementary scale it is important to bear in mind the ecological benefits derived from the hedgerows and lines of trees that separate fields in areas such as Bosc de Tosca, the basaltic plateau of Batet de la Serra and the plain of Sant Miquel del Corb. These considerations can also be applied to rivers such as El Ser and streams such as Riera de Joanetes, Riera de Sant Iscle, Riera de Cogolls, El Gurn and El Torrent de Bastons, amongst others.

Other areas of natural interest included in the *Catalogue of Sites of Natural Interest in the Counties of Girona* (Diputació de Girona, 2009) that function as natural corridors include the Vall de Bianya; Vall de Riudaura; La Miana; the basin of the river Borró and Riera de Buranc; the Ser valley; the forests of Sant Martí de Sacalm; Serra de Cogolls, Les Encies and Siubès; the headwaters of El Ritort and Riera de Mieres; and the valleys of Sant Miquel de Campmajor and the river Rodeja.

Fig. 1.9 Basaltic cliff at Castellfollit de la Roca. *Source* Catalan Landscape Observatory

Along with the legacy of its volcanoes, the harmonious distribution of croplands, pastures, forests and rural farmsteads resulting from the use of wise agropecuary and forest techniques has contributed to the many aesthetic values present in the Valls d'Olot landscape unit. These values are enhanced by the organised, balanced serenity of the sites in which this combination of different land uses is present. The contrasts in the landscape and its plasticity of forms, colours and textures are an important feature from an aesthetic perspective. The vestiges of the volcanic activity, with morphologies created by craters, cones and the prismatic columns of the basaltic lava flows, ensure that the patterns present in this landscape are unlike any other in the Iberian Peninsula. The combination of the arable land surrounding the forested volcanic cones, the crop rotation, the sinuous terraces that follow the contours of the land, and the grassy crater bottoms has evolved into a landscape of great beauty.

Of note too is the aesthetic and visual effect of the urban architecture of the village of Castellfollit de la Roca, perched high upon the areas's largest basaltic cliff and compressed vertically between the rivers Fluvià and Turonell (Fig. 1.9). This is one of the most published of all promotional images of the region, despite the construction of the nearby road viaduct on the A-26. Other visually singular sites include the historical centre of Besalú and the work carried out to restore the quarry in the volcano of Croscat, winner of a FAD prize in 1994, which has converted this site into one of the main visitor attractions in the volcanic zone.

The mosaic of fields, forests and pastures, complemented by the volcanic relief and enriched by human constructions including farms, churches and walls of volcanic stone, has generated incomparable images such as the aforementioned basaltic plateau of Batet de la Serra, a wonderful example of an agropecuary landscape with scattered settlements in the verdant heart of Catalonia.

No less appealing is the view of the Vall d'en Bas from above; its agricultural plain is criss-crossed by the geometrically straight lines of its field boundaries and a regular grid of roads, and by the incessant meandering of the course of river Fluvià insinuated by the riparian vegetation that

Fig. 1.10 Tree-lined street in the Malagrida New Town. *Source* Catalan Landscape Observatory

extends along its banks. The visual effect is intensified by the valley's hamlets and their attractive bell towers, narrow streets and balconies, as in El Mallol; especially well known are the balconies along Carrer Teixeda in the village of Els Hostalets, which are decorated with drying corn cobs and geraniums pots that lend this village a particularly harmonic and bright air. This is one of the best-known and most-photographed streets in the Valls d'Olot; at the end of the street, the horizontal sense is broken by the bell tower of the church of Santa Maria with its pyramidal roof, all with the mountains of Puigscalm, Santa Magdalena and Els Plans de Falgars as a backdrop.

The aesthetic dimension is unmistakeable in places such as La Moixina, one of the most humid spots in the whole area. Here, part of the lava flow emitted by the volcano Croscat that reached the river Fluvià, sunk in the soft sediments of the valley bottom, thereby giving rise to a landscape in which the volcanic rocks in some places emerge above the sediments and the watercourses that abound here. Numerous springs gush from the ground, in company with drainage channels, patches of marshy woodland and dry-stone walls that help overcome the slopes of the hillsides.

Human intervention in the landscape has been equally successful in certain sites in the city of Olot such as the Plaça Clarà, the Malagrida New Town (Fig. 1.10), Passeig de Barcelona and Parc Nou. This latter park (just over 3 ha) is the embodiment of a symbiosis between an urban park, a botanical garden and a fragment of woodland of great environmental value due to the plant species that grow there.

The chromatic changes deriving, above all, from plant life cycles affect aesthetic considerations. These effects are patent in places such as the D'en Jordà beech forest, which changes in colour from intense green in spring and summer to ochre and brown in autumn, before losing its leaves totally in winter.

In fluvial environments aesthetic effects are multiplied by the interplay between the water and the bountiful waterside vegetation. Aside from natural waterfalls and pools, the old mill weirs have created a number of water surfaces that local people often use as swimming pools in summer. Good examples of waterfalls include the Salt dels Portuguesos, Salt del Sallent in the Vall d'en Bas and the Salt del Molí dels Murris near Sant Feliu de Pallerols, while some of the most popular swimming holes are at Pla d'en Xurri on the river Gurn upstream from Sant Privat d'en Bas, and at Tussols-Basil in Olot.

Over the centuries, human activity and intervention in the territory have conferred a historical component on the landscape. The heritage deriving from the constructions built from local prime materials has created a series of recognisable elements in the landscape. Volcanic rock has been used for centuries to build houses and farms, shepherds' huts and dry-stone walls (without mortar). The main areas of dry-stone construction are found around Olot (and, above all, Les Preses) on top of the lava flows. These landscapes are characterized by walls and piles of volcanic rocks that divide the agricultural land into small plots. There are also numerous huts, most encrusted into the dry-stone walls. The two sectors of the Valls d'Olot where the dry-stone constructions are visually most stimulating are the basaltic plateau of Batet, and the lava flow in the Bosc de Tosca between Olot and Les Preses. The size, colours, shapes and textures of the fragments of basaltic rock vary according to their origin.

Further afield, the area also contains a number of monumental buildings of great architectural value built from volcanic rock, which are obligatory points of reference in the local landscape. Of note is the old village of Santa Pau, declared of historico-artistic interest in 1971. Standing on a rocky outcrop next to the river Ser, the village has a series of medieval elements including an old encircling defensive wall, which protects the houses inside that huddle around the thirteenth-century castle. Also of note here is the fifteenth-century church of Santa Maria, with its robust bell tower, and the muddle of narrow streets with sudden bends and steep slopes that all lead to one of the old entrances into the walled village, the Portal del Mar. In order to control the relationship between the old village and the urban and natural areas that surround it, in 2000 the Catalan government established a buffer protection zone that includes the whole of the outcrop on which the village stands.

Lying in the heart of the agricultural plain of the Vall d'en Bas, the village of Els Hostalets d'en Bas has also been declared of historico-artistic interest. Its value lies in the uniformity of its buildings and the materials used in its construction. Nearby, also in its heart, stands the once most important settlement in the valley, the village of El Mallol. Built on a 70-m high outcrop, this village's splendid architectural styles can still be admired, despite the destruction of the fortress that was once part of the village.

Besalú, one of the most interesting and best-preserved medieval sites in Catalonia, possesses a magnificent built heritage whose splendour is patent throughout its streets and in its civil and religious buildings.

Other architectural sites that should not be overlooked include the village of Sant Privat d'en Bas, whose medieval centre is reached through an arched passageway; Joanetes, constructed on a hill on the slopes above the Vall d'en Bas, with the unmistakeable profile of the mountain of Puigsacalm acting as a backdrop; Sant Feliu de Pallerols, bisected by the river Brugent, and, of course, the old centre of Castellfollit de la Roca whose houses crowd together along the narrow edge of the basaltic cliff standing between the rivers Fluvià and Turonell.

A separate case is that of the city of Olot, whose urban centre conserves no large buildings from before the fifteenth century due to the earthquakes that hit the city in 1427 and 1428. The reconstruction of the city allowed a new settlement with a more modern centre to be constructed around the main square, La Plaça Major. The parish church of Sant Esteve (eighteenth century) stands out vigorously from the urban landscape but, in truth, it is the church of La Mare de Déu del Tura that is most closely linked to the birth of the city. Also of interest here are the Renaissance cloisters of El Carme, declared of national cultural interest and current home to the Olot School of Art. Typical corners in the urban panorama of Olot include the Plaça Clarà, the Malagrida New Town (whose morphology is very reminiscent of the English garden cities), the banks of the river Fluvià, Passeig de Barcelona and the aforementioned Parc Nou.

From an architectural point of view, it would be erroneous to ignore the scattered buildings—from the simple churches in small rural settlements to the farmsteads in the plains and in the mountains—that dot the landscape of the Valls d'Olot. A succession of churches, chapels, shrines, small fortifications and watchtowers occupying fine vantage points dominate great swathes of land and are fully fledged and essential parts of the image of the landscape of the Valls d'Olot. The best such examples are the church of Santa Margarida in the centre of the crater of the volcano of the same name; the church of Sant Francesc and the fortifications on the top of the volcano of El Montsacopa; Sant Miquel del Mont on top of the homonymous mountain ridge, and the Sanctuary of El Cós.

Other elements that have enriched the landscape from a historical perspective include the pre-industrial and industrial hydraulic installations along the river Fluvià (e.g. textile factories, mills, mini-HEP plants, weirs, canals, ponds and bridges), above all on the stretch of river between Olot and Castellfollit de la Roca.

Finally, it is worth highlighting the zigzags described by the Via Annia Roman road, a branch of the Via Augusta that was used right up to the beginning of the twentieth century as a natural pass over Capsacosta to link the counties of La Garrotxa and El Ripollès. Currently, this Roman road is signposted as a footpath between Sant Salvador de Bianya and Sant Pau de Segúries.

The rich landscape of the Valls d'Olot cannot be understood without taking into account the productive activities practiced over the centuries that confer a special significance on particular sites and areas. Agropecuary exploitations have been—and still are—fundamental in shaping the landscape and have had the most direct effect in the Vall d'en Bas, an agrarian area that boasts some of the richest soils in Catalonia. The yields of the main crops—maize and forage species—are high and production takes place in a landscape characterised by regular-sized fields separated by a rectangular network of roads, the fruit of the land consolidation practiced in 1966–1972. Elsewhere, the Vall de Bianya is a similar area and shares a number of agrarian particularities with the Vall d'en Bas.

Santa Pau haricot beans are one of the best-known local products from the Valls d'Olot and their cultivation goes back a long way in time. Part of the production of these beans is sold under the brand name *Fesols de Santa Pau* by an association of producers and retailers.

The gastronomic group *Volcanic Cuisine* was created in 1994 as a means of promoting agrarian products from the Valls d'Olot in gastronomic circles and as a tourist attraction. This brand identifies dishes produced with some of the most singular products of the region: *escarlots* (a wild mushroom), Santa Pau haricot beans, buckwheat, turnips, chestnuts, potatoes, maize, snails, wild boar and pork.

Economic activities such as agro-alimentary production, gastronomy, accommodation services and leisure activities all benefit from the presence of a high-quality landscape, as do the quality brand names and labels that have been created to distinguish certain local products.

Tourism reaps its rewards from the unquestionable lure of the local landscape, which acts as a first-grade attraction for many different types of visitors. Walking has long been practiced in the area and in recent years new ways of enjoying the landscape have emerged: families come in groups for long stays in hotels or second homes, while pensioner groups from abroad tend to visit for shorter periods of time. Scientific and educational tourism related with the volcanic zone is also burgeoning, as is the tourism attracted by area's artistic tradition. Rural tourism has grown as people restore old farmhouses and town houses for rent as independent accommodation. Companies dedicated to guiding walkers, horse-riding and even hot-air-balloon flights have all sprung up, while the Girona-Olot greenway cycle path that follows the route of the former railway line has stimulated cycling-based tourism.

Another productive sphere is the generation of HEP, which still occurs on a small scale along the banks of the river Fluvià between Olot and Sant Jaume de Llierca.

The social values of the landscape manifest themselves in a number of ways. In recent years a large number of walking routes have been signposted that enable visitors and local people alike to discover the landscapes of the Valls d'Olot in a respectful, sustainable fashion. The existence of La Garrotxa Volcanic Zone Natural Park has played a decisive role in this trend. More recently, *Itinerànnia*, a network of signposted paths and trails that links the counties of L'Alt Empordà, La Garrotxa and El Ripollès, has been set up and now links practically all the villages and towns in the area. In addition, the GR-2 and GR-83 long-distance footpaths run through the area and many people make good use of the cycle path that follows the route of the railway line that once ran from Girona to Olot. This trail is used as much for sport as for social and cultural reasons and has become very successful as a space for gentle walking, either individually or in groups (Fig. 1.11).

Some of the region's most striking landscapes are in fact those that have been designed to attract visitors and/or to encourage people to fully appreciate their surroundings. Recreational areas with tables, car-parks and short signposted walks exist, along with an increasing number of excellent look-out points equipped with interpretation panels. The best known such areas are the Àrea Recreativa de Xenacs in Les Preses and Pla d'en Xurri in Sant Privat d'en Bas. In Olot, the Àrea Recreativa Tussols-Basil on the banks of the river Fluvià and La Moixina also function as leisure areas.

One of the developments that has most improved the landscape is the restoration of the locally well-known and well-used Parc de Pedra Tosca in Les Preses, which was funded by a European Union LIFE project. The project, finished in 2004, was awarded the IV Rosa Barba Landscape Prize.

The towns of Olot, Les Planes d'Hostoles, Castellfollit de la Roca and Sant Jaume de Llierca all have large areas of semi-urban allotments that play an important productive, social, cultural, environmental and urban role in local communities.

Finally, it is worth mentioning the markets and fairs that are held regularly in many of the towns and villages in the Valls d'Olot.

The singularity and richness of the landscape of the Valls d'Olot guarantees their symbolic and identitary value, and act as an allure for those wishing to come to perceive new experiences. Places such as the crater of the volcano of Santa Margarida, with its small church in its centre; the basaltic cliff at Castellfollit de la Roca, crowned by a line of houses; Carrer Teixeda in Els Hostalets d'en Bas; the volcanoes of Croscat and El Montsacopa; the castle and medieval centre of the village of Santa Pau; the D'en Jordà beech forest; Besalú and its medieval bridge; and the marshy woodland of La Moixina are all sites that are immediately identifiable as essential elements of the Valls d'Olot.

Beyond the large parish churches, the religious architecture of the area is embodied by many small isolated churches and chapels in rural areas. This succession of buildings greatly enriches our appreciation of the landscape, even more so since, in the case of the Valls d'Olot, many were

Fig. 1.11 Leisure activities around Olot. *Source* Catalan Landscape Observatory

erected in places that offer excellent panoramic views over the landscape.

The largest and the best known of the volcanic craters look out over the surrounding flat or rolling land and enhance the unmistakeable silhouettes of their volcanic edifices. This effect can be appreciated from the vantage points at Xenacs and Colltort, from where there are excellent perspectives over the surrounding volcanoes and the D'en Jordà beech forest. No less striking is the contrast—obvious from the hilltop village of El Mallol—between the flat Vall d'en Bas and the sheer cliffs of the mountains of Santa Magdalena-Puigsacalm, Els Llancers and Falgars that tower above it, above all to the west.

Certain landscapes are indelibly connected to events of the past that are still, to a lesser or greater extent, evoked by current generations. Places such as the Verntallat house or the ruins of the castle of Hostoles, built on a rocky outcrop overlooking the Brugent valley, have an important symbolic content due to the part they played in the wars of the Catalan Peasant Revolt in the fifteenth century.

During the nineteenth-century Carline Wars, the mountainous and forested Valls d'Olot were the scene of a number of armed skirmishes. The most suggestive of places in the area is, possibly, the perfectly preserved Hostal de la Corda, situated between the municipalities of Olot and Riudaura. There, on 26 March 1875, the day after the Carline troops had been ousted from the city of Olot, a secret interview was held between generals Arsenio Martínez Campos, head of the Liberal forces, and Francesc Savalls, head of the Carline troops in Catalonia, that was to signal shortly the end of the Carline resistance in Catalonia.

The tradition of designing and building of nativity scenes has been well established in the region since the nineteenth century; this practice captures the idiosyncrasies of the landscape of the Valls d'Olot and its principal characteristic is its faithfulness to the local landscapes. The mountains and cliffs, the valleys, the judiciously distributed farmsteads, and the churches strategically situated on hilltops and in places of great beauty, provide a backdrop to many of the nativity scenes that are constructed every year—further evidence of the close and deep-rooted connection that exists in the region between artistic and cultural activities and the landscape.

1.2.5 Main Routes and Places from Which to Enjoy Views of the Landscape

The main routes and vantage points from which to observe the landscape constitute part of the character of that landscape. Such places are accessible, offer a broad vision of the landscape, and are attractive and varied; they enable the viewer to capture all the shades and nuances of the landscape at different scales, appreciate its inherent values and dynamics, and even interact with it if need be. Thus, these special sites are not merely places for visual contemplation as they also offer a sensorial, emotional and living experience. The Valls d'Olot possess a whole series of wonderful observation points, although, given that most roads stick to the lower parts of the valleys, it is often necessary to walk for a varying distance to be able to enjoy the best views in full.

Of the roads that run through the area, the GI-524 from Banyoles to Olot via Mieres and Santa Pau is perhaps the most interesting as it passes through a succession of vegetation zones, ranging from the lower Mediterranean forests to the more humid woodland in the Olot basin.

The road known as the 'Parcel·laria', built when the agricultural land of the Vall d'en Bas was being consolidated, is a quiet back road that runs from one end of this valley to the other (Fig. 1.12). Along with the lanes that adjoin it, this road offers visitors the chance to visit some of the most attractive landscapes in the valley, and to discover its agricultural history via the farmsteads and hamlets that dot the landscape.

The GIV-5241 gives access to the plateau of Batet de la Serra and hamlets such as La Trinitat. It connects to a tarmacked track heading towards Santa Pau that runs close to

Fig. 1.12 One of the minor roads that cross the Vall d'en Bas. *Source* Catalan Landscape Observatory

the volcanoes of Puig de Pujalós, Puig Safont and Puig de Martinyà, through the mosaic of fields and copses that characterise the Batet plateau. Another possibility is to discover the Vall de Bianya along the minor road that connects Llocalou to the hamlets of Capsec and Sant Pere Despuig.

For walkers, the long-distance footpaths such as GR-2 and GR-83 that cross the Valls d'Olot in numerous different directions are of great use. The former runs east-to-west, from Besalú to Sant Miquel de Castelló via Santa Pau, while the latter connects Mataró and the mountain of El Canigó via sites in the Valls d'Olot such as Cogolls, Fontpobra and Santa Margarida. Additionally, GR-1 passes through Besalú. The creation of La Garrotxa Volcanic Zone Natural Park has led to the signposting of 28 local footpaths that take walkers to the most outstanding sites in the park. Good examples include the route through the valleys of Sant Iscle and El Vallac, the itinerary that links the D'en Jordà beech forest and the volcanoes of Santa Margarida and Croscat, and the trail that takes visitors to the three lava flows and basalt cliffs in Sant Joan les Fonts. Last but not least, the greenway for cyclists that links Olot and Girona is another excellent way of getting to know the landscapes in the Valls d'Olot.

Many varied vantage points—of which a number can be reached along the signposted walks in the Natural Park—provide spectacular views over the Valls d'Olot and reveal the exceptionality of this humanised volcanic landscape.

The viewpoint at Puig Redon is one of the most attractive features of the Àrea Recreativa de Xenacs (Les Preses) and offers uninterrupted views northwards over the Olot basin, Croscat and the D'en Jordà beech forest. Likewise, there are also excellent views—complemented by an interpretation board—of the whole amphitheatre of mountains running from Puigsacalm in the west to the Pyrenees in the sector Núria-Ulldeter, El Canigó and L'Alta Garrotxa in the east.

The volcano of El Montsacopa as a vantage point is also highly recommendable. From the walk that circumnavigates the crater lip, there is a wonderful view of the whole of Olot, as well as of the main volcanoes and ridges that encircle the city. The castle of Finestres—somewhat less accessible than the other viewpoints—also offers a full view of the landscape of the Valls d'Olot, as well as good perspectives over the Llémena valley and the Mieres basin.

Although not strictly part of the Valls d'Olot, the following sites also provide exceptional views over these valleys. Sant Miquel de Castelló (also known as Falgars) offers a complete if somewhat vertiginous view over the Vall d'en Bas, while those who climb to the summit of Puigsacalm will be rewarded with a 360° panorama, the most complete of all views of the Valls d'Olot. The viewpoint (with interpretation panel) at the Mare de Déu de la Font de la Salut in Sant Feliu de Pallerols in the Cabrerès-Puigsacalm landscape unit is the best in the southernmost sector of the Valls d'Olot, providing, above all, a striking vision of the Hostoles valley.

1.2.6 Objectives of Landscape Quality

The drafting of *Catalogue of the Landscapes of the Counties of Girona* outlined the following eight quality objectives for the area, which were defined after consultation with the main agents and entities that work on the landscape in this region.

- La Garrotxa Volcanic Zone Natural Park: well-conserved, ecologically viable and helping to ensure that economic activities, above all in the agropecuary sector, are compatible with the use of natural resources, tourism and other leisure activities.
- The towns and villages of Castellfollit de la Roca, Mieres, Les Planes d'Hostoles, Les Preses, Olot, Sant Feliu de Pallerols, Sant Jaume de Llierca, Sant Joan les Fonts, Argelaguer, Besalú and Santa Pau: well run and in harmony with the features of the landscape that define them and the surrounding land; attractive entrances to their urban centres.
- Communication networks and their surroundings that are well integrated into the landscape and do not fragment current systems of territorial interconnection.
- Special-use areas (industrial, logistic, commercial, leisure and other tertiary services) in Les Preses, Olot, Sant Joan les Fonts, Sant Esteve d'en Bas, Les Planes d'Hostoles, Sant Feliu de Pallerols, Sant Jaume de Llierca, Vall de Bianya and Besalú to be located in visually non-intrusive areas and designed bearing in mind their integration into the surrounding landscape.
- The river Fluvià in the Valls d'Olot: well conserved in terms of the quality and quantity of its waters, lined by well-preserved riparian woodland that can act as a landscape corridor and as a space for leisure activities.
- A productive agricultural landscape, well preserved and properly managed, that conserves the many varied features that characterise it and define its personality, above all in the unique combination of fields, pastures, allotments, farmsteads, woodland and hedgerows.
- Landscape values preserved (Fig. 1.13) and valued as part of local heritage, and quality visual reference points that guarantee the survival of a local sense of identity.
- A system of routes and viewpoints that embrace the most important local panoramas to allow people to discover and interact with the diversity and nuances of the landscapes of the Valls d'Olot.

Fig. 1.13 Restoring the landscape inside the volcano Croscat. *Source* Catalan Landscape Observatory

1.3 Characteristics of the Volcanic Landscape Beyond the Valls d'Olot

As commented in the previous section, aside from the Valls d'Olot, a small part of La Garrotxa volcanic zone coincides with the landscape units of Rocacorba and La Plana de la Selva. The volcanoes of Banya del Boc, Puig Montner, Clot de l'Omera and Puig d'Adri are some of the most striking features of the landscape of Rocacorba, while in the Plana de la Selva, the volcanic activity has left its mark in the form of the singular landscapes of the volcano of Crosa de Sant Dalmai, the volcanic hills of Maçanet and the thermal springs at Caldes de Malavella and Santa Coloma de Farners.

1.3.1 The Volcanic Landscape of Rocacorba

The Rocacorba landscape unit covers 272 km^2 and encompasses a whole series of ridges and valleys running predominantly north-south in a roughly square area bounded to the north by Serra de Corb and Serra de Finestres (outside this landscape unit), to the west by the Brugent valley, to the east by the basin of Lake Banyoles, and to the south by the Ter valley (Fig. 1.14). Its ceiling is in the mountains of Rocacorba, where the peak of Puigsou reaches 992 m a.s.l. Aside from the areas of more abrupt relief marked by imposing vertical cliffs, the land surface of this landscape unit is scored by many small valleys of tectonic origin and

Fig. 1.14 The mountains of Rocacorba, with the two antennae on the summit of its principal landscape feature. *Source* Catalan Landscape Observatory

possesses a network of rivers and valleys—Brugent, Llémena and Sant Miquel de Campmajor—where most of the local agricultural activity and population are concentrated.

The Rocacorba landscape unit is home to volcanic edifices such as Puig de la Banya del Boc, El Clot de l'Omera, Puig Montner and Puig d'Adri. These singular morphological features, whose intrinsic value is recognised by their inclusion in the *Catalogue of Areas of Geological Interest*, are true visual focal points that possess a great symbolic and identitary value for the local population.

Over three-quarters of this area is covered by vegetation, including extensive holm-oak and, to a lesser extent, cork-oak forests. Forests thrive on the steepest, most abrupt slopes, above all in highest parts of the northern and eastern sectors of the area, where the landscape has become fairly homogenous due to forest encroachment into areas once occupied by arable and pasture land.

A stable rural population survives in the valleys and plains of the Rocacorba landscape unit. Agriculture is mainly represented by herbaceous crops cultivated, in particular, in the valley bottoms of its main rivers, the Brugent, Llémena and Ter, and in the plains around Adri.

Built-up areas in this unit occupy very little land (around 1 %) and are split between village and town centres, out-of-town residential complexes and industrial and commercial estates. One of the main features of Rocacorba is a settlement pattern consisting of scattered farmsteads and hamlets, fruit of the repopulation in the High Middle Ages but today in many cases uninhabited. Many of the municipalities in this area possess more than one settlement of a certain size, along with numerous hamlets and farmsteads; this is the case of Sant Miquel de Campmajor, a village that encompasses a number of different parishes and many scattered farms and homesteads.

Rocacorba as a landscape is in essence orderly, harmonious and balanced, the product of a low-intensity anthropic transformation that is patent in any global vision of the area.

1.3.2 The Volcanic Landscape of the Plana de La Selva

The Plana de la Selva landscape unit corresponds to the far north of the Depressió Prelitoral and, specifically, consists of a sunken block bounded by the mountains of Les Gavarres (to the east), Les Guilleries (west) and L'Ardenya (south) (Fig. 1.15). Northwards, the Plana de la Selva merges into the Pla de Girona, with which it shares both its tectonic origin and a number of geomorphological characteristics.

The tectonic depression that gave rise to this plain has been infilled by sedimentary Neogene and Quaternary rocks (mudstones, clays, gravels, arkoses, clays and conglomerates) deposited during the Pliocene and the Upper Pleistocene. Also present are basaltic lava flows and small volcanic edifices generated by the volcanic activity that took place in the Tertiary and Quaternary ages. The most conspicuous example is the volcano of La Crosa de Sant Dalmai in the northeast of the plain. The edifice of this volcano is

Fig. 1.15 Panoramic view of the landscape of the Plana de la Selva from the slopes of the volcano of La Crosa de Sant Dalmai. *Source* Catalan Landscape Observatory

perfectly preserved and consists of a large circular crater (diameter: 1250 m), the largest example of explosive volcanic activity in Catalonia. The volcanic past and the singularity of some if its wetland habitats are the main reasons that justify the protection of many of the natural sites in the Plana de la Selva. The presence of the thermal baths at Caldes de Malavella and Santa Coloma de Farners can be explained by the system of fractures that affect the region's basement rocks and also permit the existence of the area's volcanic features.

Today's landscape in the Plana de la Selva is the product of the successive human transformations that have been imposed on the territory over historical time. Since medieval days this area has always been essentially agricultural, and arable farming and animal husbandry—as well as the exploitation of the forest patches—once represented the main human disturbance of the land surface. However, in the second half of the twentieth century new elements have severely transformed the landscape: on the one hand, out-of-town residential complexes have expanded into forest areas, while, on the other, the area has witnessed a rapid growth in communication infrastructures such as the AP-7 motorway, the widening of the N-II and the construction of a very-high-tension power line.

The plain is covered by a complex agroforestal mosaic in which herbaceous crops, plane and poplar plantations, a series of small streams and gullies, and a extensive network of tracks and minor roads fill the landscape. The villages of Sant Dalmai and Sant Andreu Salou, fully integrated into their immediate surroundings, are of note due to their singular morphologies and for having preserved their characteristic urban structures. In the distance, the silhouette of the town of Llagostera stands out from afar, visibly preeminent on its hilltop rising above the encircling plains.

2 Geological Setting of La Garrotxa Volcanic Field

Joan Martí, Xavier de Bolós and Llorenç Planagumà

Situated in the northeast of the Iberian Peninsula, La Garrotxa Volcanic Field is part of the Catalan Volcanic Zone and one of the provinces of the Neogene-Quaternary alkaline volcanism associated with the European Rift System. It covers about 600 km^2 and lies between the cities of Olot and Girona (Fig. 2.1). This basaltic volcanic field contains over 50 cones (including both cinder and scoria cones), lava flows, tuff rings and maars dating from the Middle Pleistocene to the early Holocene, which rest either on upper Palaeozoic granites and schists or on sedimentary Eocene and Quaternary substrata. Available petrological and geochemical data indicate that this region consists of a suite of intracontinental leucite, basanites, nepheline basanites and alkali olivine basalts, which in most cases represent primary or near-primary magmas, their geochemical characteristics being very similar to analogous petrologic types found in other European Cenozoic volcanic zones.

La Garrotxa Volcanic Field embraces two geographically distinct zones, the larger area located in the north of the county of La Garrotxa, mostly corresponding to La Garrotxa Volcanic Zone Natural Park, and a southerly area that contains fewer but larger and more complex volcanic edifices (Fig. 2.1). Although both correspond to tectonically controlled depressions, the northern zone has substrata consisting of thick layers of Tertiary and Quaternary sediments, whereas the southern zone is underlain by unconsolidated Quaternary sediments in combination with the Palaeozoic basement.

Volcanic activity in La Garrotxa Volcanic Field is characterised by numerous small cinder cones built during short-lived monogenetic eruptions occurring along tectonic-related volcanic fissures. The total volume of extruded magma in each eruption was between 0.01 and 0.2 km^3 (DRE). Strombolian and phreatomagmatic episodes alternated in most of these eruptions and gave rise to complex stratigraphic sequences with a broad range of pyroclastic deposits. The eruption sequences differ from one cone to another and demonstrate that the eruptions did not follow a common pattern, particularly in cases of magma/water interaction. This complex eruptive behaviour is likely to be due to the differing stratigraphic, structural and hydrogeological characteristics of the substrata below each volcano rather than to any differences in the physicochemistry of the erupting magmas, which are generally fairly homogeneous throughout La Garrotxa Volcanic Field.

The existence of this volcanism is linked to the complex geodynamic evolution of the area following the Alpine orogeny that involved great stretching and breakage of the continental lithosphere, thereby allowing the generation of mafic magmas in the mantle and their subsequent ascent and eruption. The evolution of La Garrotxa Volcanic Field is chiefly controlled by two major Neogene faults, the Amer and Llorà faults, oriented NW-SE like most of the major post-Alpine extensional faults that have defined horst and graben structural patterns in NE Iberia. However, most of the eruptive fissures and secondary structural lineaments that control the volcanic activity in La Garrotxa Volcanic Field exhibit a NNW-SSE trend that runs slightly obliquely to the main faults.

The volcanic activity in La Garrotxa Volcanic Field occasioned the accumulation of thick layers of volcanic rocks that, in combination with the particular microclimate of the area, has guaranteed the formation of fertile soils covered by dense vegetation, a process that has helped preserve some of the original volcanic morphologies.

J. Martí (✉) · X. de Bolós
Institute of Earth Sciences Jaume Almera CSIC, Barcelona, Spain
e-mail: joan.marti@ictja.csic.es

X. de Bolós
e-mail: xavier.bolos@gmail.com

L. Planagumà
Tosca, Environmental Services, Olot, Spain
e-mail: llplanaguma@gmail.com

J. Martí and L. Planagumà (eds.), *La Garrotxa Volcanic Field of Northeast Spain*,
Geoheritage, Geoparks and Geotourism, DOI 10.1007/978-3-319-42080-6_2

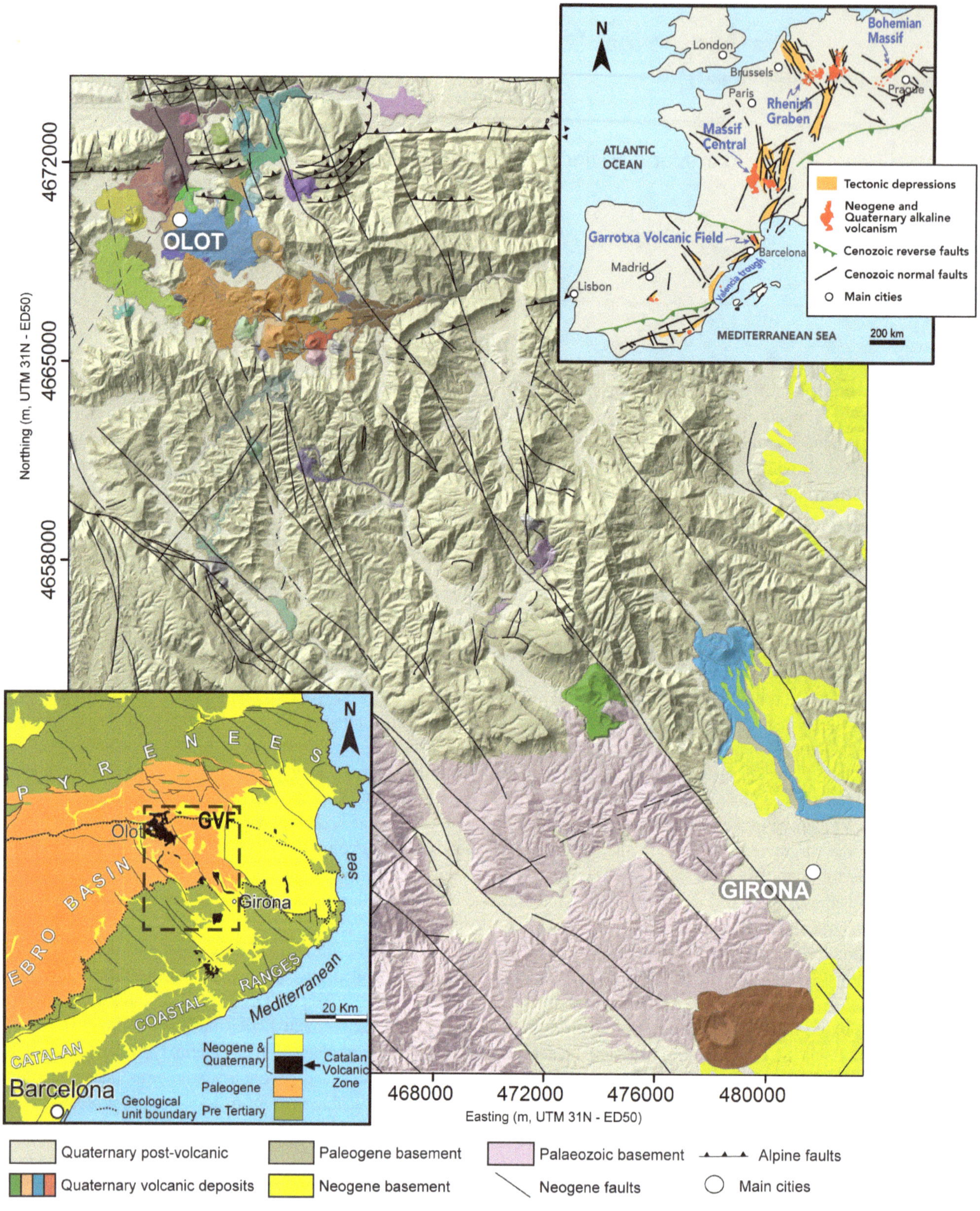

Fig. 2.1 Geological setting the study area. Inlet at the *upper right corner* structural map of the European Rift System. Inlet at the *lower left corner* geological map of the Ne of Spain. *Central image* geological map of La Garrotxa Volcanic Field

2.1 Geodynamic Context

The European Cenozoic Rift System extends from the coast of the North Sea to the Mediterranean and consists of the Spanish Valencia Trough, the Gulf of Lion and Massif Central in France, the Rhine, Ruhr Valley and Leine grabens in Germany, which straddle the river Rhine and cut across the Rhenish Shield, and the Eger Graben in the Bohemian Massif (Fig. 2.1). In the south it joins a Plio-Pleistocene volcanic chain crossing the Atlas ranges. This rift system evolved in the Alpine foreland during late Eocene to Recent times. The evolution of this European rift system is thought to be governed by the interaction of the Eurasian and African plates and by early phases of a plate-boundary reorganisation that may lead to the break-up of the present assembly of continents.

Widely distributed along an extensive rift system, the Cenozoic (Middle Miocene to present) alkaline volcanism in central and western Europe consists of four main volcanic areas: the Rhenohercynian (Germany), the western Panonian Basin (Eastern Europe), the Massif Central (France) and the Valencia Trough (Spain). The causal mechanism(s) of this rift system is poorly understood but seems to be related to the extension that affected the whole area following the Alpine tectonics, which was probably associated with the upwelling of a mantle plume beneath the European plate. This extension would thus explain the creation of a number of tectonic basins in which Neogene-Quaternary sedimentation and volcanism have occurred. Primitive mafic alkaline volcanic rocks derived from this volcanism have petrological and geochemical signatures that suggest the involvement of both lithospheric and asthenospheric mantle source components in their petrogenesis.

The Valencia Trough is a NE-SW-oriented Neogene basin located between the Iberian Peninsula and the Balearic archipelago offshore of NE Spain (Fig. 2.1). It has a complex geological history that contains two main stages of magmatism. During Early to Middle Miocene times, the area was subjected to compressional tectonics accompanied by calc-alkaline volcanism. This was followed by a period of extensional tectonics and mafic alkaline volcanism in the Middle Miocene to Recent times. The greatest concentration of Middle Miocene to Recent volcanism in the region is found in the Catalan Volcanic Zone (CVZ) in the NE Iberian Peninsula (Fig. 2.1), in which La Garrotxa Volcanic Field represents the most recent (0.7 Ma to early Holocene) episodes of this volcanism (Fig. 2.1).

The CVZ has traditionally been divided into three sub-zones based on the age of its volcanic rocks: L'Empordà in the northeast (>12–8 Ma), La Selva (7.9–1.7 Ma) in the south and La Garrotxa (>0.5–0.01 Ma) in the west (Fig. 2.1). The total volume of extruded magma seems to increase progressively from the early (L'Empordà) to later (La Garrotxa) episodes. Thus, a progressive and concomitant increase in the volume of magma generated, as well as an increase in the degree of partial melting, can be observed in the geochemistry of the rocks in the CVZ. Some volcanoes in La Garrotxa sub-zone contain ultramafic to mafic xenoliths. The xenoliths comprise pyroxenites (the most abundant), melanogabros, amphibolites and spinel lherzolites. Pressure and temperature estimates for these xenoliths suggest that they may have crystallised in magma chambers located at the crust-mantle boundary, which geophysical estimates locate at a depth ~30 km. These geophysical studies also indicate that the CVZ is characterised by a regionally thinned lithosphere, about 60–70 km thick, high elevations and a notable thermal gradient, suggesting that the area is affected not only by the topographic load of the Pyrenees but also by the opening of the Valencia trough. The local structure of the area is composed of a set of horst and grabens bounded by NW- and NE-oriented Neogene normal faults that have controlled the recent sedimentation and distribution of the area's volcanism.

2.2 Geological Characteristics of La Garrotxa Volcanic Field

The CVZ, which includes La Garrotxa Volcanic Field, is located in the NE corner of the Iberian Peninsula, and is limited by the eastern Pyrenees (north), the Ebro basin (west) and the Catalan Coastal Ranges (south) (Fig. 2.1). Its geological evolution is complex and includes the formation of a Palaeozoic basement, highly deformed by the Variscan orogeny, the sedimentation of a thick sequence of Mesozoic and Tertiary rocks, folding and faulting during the Alpine orogeny, and, finally, the Neogene-Quaternary extension that has controlled recent sedimentation and volcanism. Consequently, the lithostratigraphic units that outcrop in La Garrotxa Volcanic Field and form the basement of the volcanic edifices correspond to materials from the upper Palaeozoic, Eocene and Quaternary ages (Fig. 3.1). As a consequence of the Alpine folding, the Neogene normal faulting system and subsequent erosion, the basement of each volcano varies. The oldest recognisable unit corresponds to the schist, gneiss, granodiorites and granites of Permo-Carboniferous age, which is unconformably overlain by the Eocene Formations that, from base to top, comprises: (1) the blue marls and gypsum of the Banyoles Formation; (2) the marls and brown sandstones of the Bracons Formation; (3) the red sandstones, mudstones and conglomerates of the Bellmunt Formation; (4) the glauconite sandstones and conglomerates of the Folgueroles Formation; and, finally, (5) the grey sandstones and marls of the Rocacorba Formation. Filling valley bottoms or the existing trough in the area and unconformably overlying the previous units lie

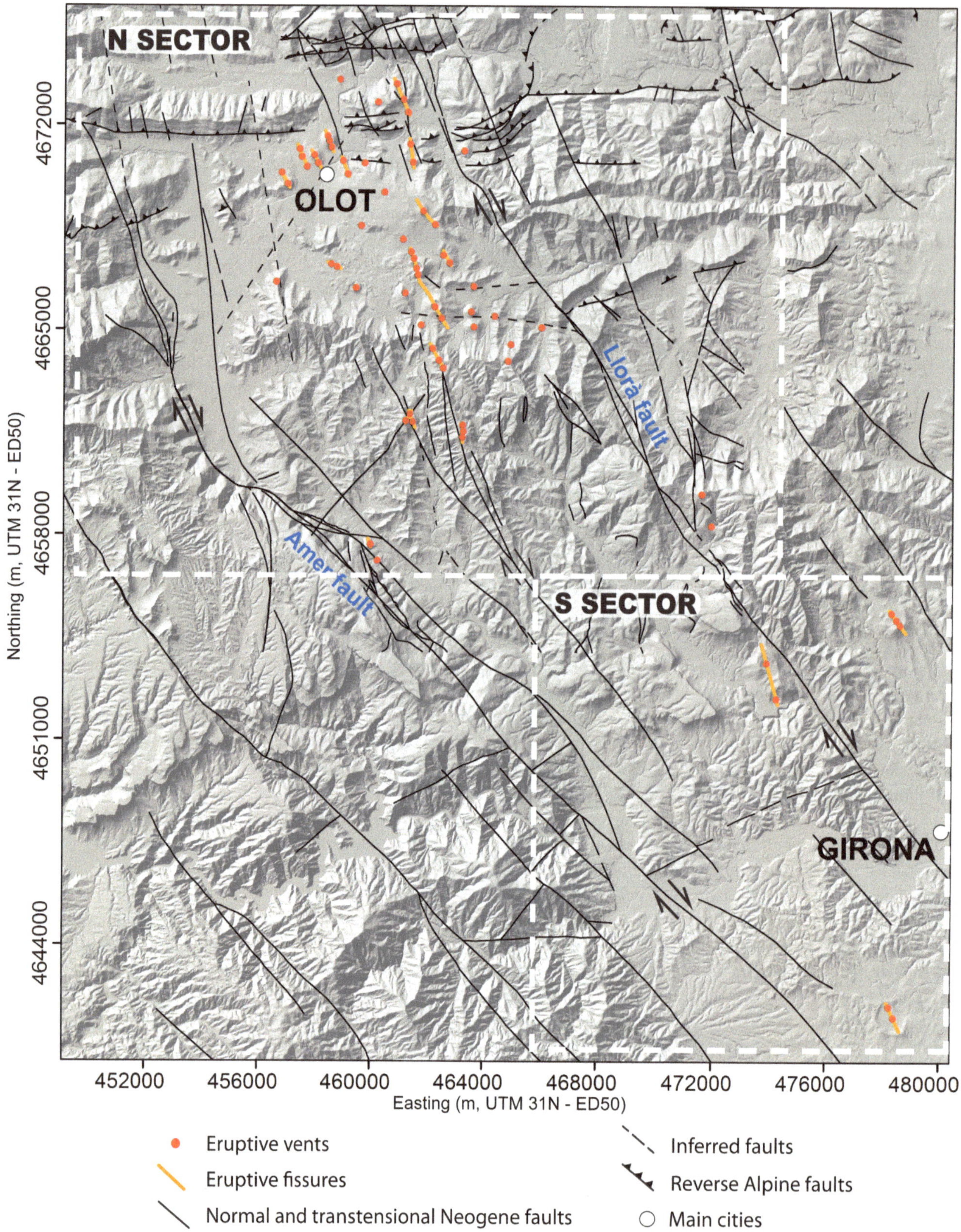

Fig. 2.2 Structural map of La Garrotxa Volcanic Field, with indication of all main structural elements including positions of vents and eruptive fissures (reproduced with permission of Elsevier)

unconsolidated gravels, clay and sands, and alluvial deposits, which, together with lava flows and pyroclastic products, form the Quaternary sequence. The Palaeozoic terranes, the Bellmunt Formation and the Quaternary deposits hold the main aquifers of the area, although the bases of the Folgueroles and Banyoles Formations may also act as aquifers in some sectors of the study area.

Due to the extensional tectonics following the Alpine compression in western Europe that were responsible for the formation of the European Rift System, NE Iberia developed a horst and grabens structure mainly limited by NE- and NW-oriented major faults. These faults show normal and trans-tensional movements and control sedimentation and volcanism. In particular, La Garrotxa Volcanic Field is bounded by two regional conjugated Neogene normal faults —the Amer Fault to the east and the Llorà Fault to the west —with a trans-tensional component. These two faults are responsible for the distribution of the area's volcanism and its seismicity, as well as the structuring of its fluvial network (Fig. 2.2).

2.3 Volcanism

The first volcanic episodes in the CVZ took place in the Empordà sub-zone and have ages older than 12 Ma. Since then, volcanism has been intermittent, producing an eruptive episode every several tens of thousands years up to the early Holocene. The age of these rocks suggests that a migration in the loci of the volcanism occurred, first southwards and subsequently towards the northwest, probably as a response to a migration in the focus of local tectonism. Thus, La Garrotxa Volcanic Field contains the most recent volcanism in the CVZ and is also the most tectonically active area, with historically high-magnitude seismicity.

Available data indicate that mafic volcanic products in the CVZ, like the parental magmas of the cumulate xenoliths, range from strongly silica-undersaturated to nearly silica-saturated compositions. This region comprises a suite of intracontinental leucite, basanites, nepheline basanites and alkali olivine basalts, which in most cases represent primary or near-primary magmas.

2.3.1 The Volcanoes of La Garrotxa Volcanic Field

La Garrotxa Volcanic Field hosts the youngest and best-preserved volcanic edifices in the whole CVZ. Over 50 volcanic cones are recognisable and can be grouped into two discrete areas, a northern sector corresponding to the upper basin of the river Fluvià and a southern sector located in the middle reaches of the basin of the river Ter (Fig. 2.2). The main concentration of volcanic cones and edifices lies in the northern sector, which corresponds to La Garrotxa Volcanic Zone Natural Park, while the southern sector holds far fewer but larger cones. The basement on which these monogenetic volcanoes stand differs between the two sectors. In the north, the volcanic rocks lie on Tertiary sediments, while towards the south they rest in some cases directly on the granites and schists of the Palaeozoic basement. During the Quaternary, volcanic activity occurred sporadically in the study area over a time period ranging from >500,000 years ago to about 11,000 years ago, with eruptive events occurring every 10,000–30,000 years.

Volcanism in La Garrotxa Volcanic Field is characterised by the presence of small cinder cones constructed during short-lived monogenetic eruptions associated with widely dispersed fractures of short lateral extent (Fig. 2.3). The total volume of extruded magma in each eruption was small (0.01–0.2 km^3 DRE), suggesting that the amount of magma available to feed each eruption was very limited. Strombolian and phreatomagmatic episodes alternated in most of these eruptions and gave rise to complex stratigraphic sequences composed of a wide range of pyroclastic deposits.

All the studied volcanoes were constructed during a single eruptive episode (i.e. they thus should be referred to as 'monogenetic') that commonly included several distinctive phases with no significant temporal separations between them. We can separate two groups of volcanic edifices: those that were built only by Strombolian activity and those that also experienced some phreatomagmatic phases. In the first case, the volcanic edifices are symmetrical or horseshoe-shaped cinder cones constructed by the accumulation of scoria and lapilli, altered by occasional emissions of lava flows. Examples of this type of activity include the volcanoes of Puigalós, Puig de Martinyà, San Marc, Roca Negra and Puig Subià (Figs. 2.4 and 2.5). Volcanic cones that experienced phreatomagmatic activity are much more complex, although morphologically they are still Strombolian in type. In these cases, the eruptive activity was characterised by a succession of phreatic phases produced by vapour explosions that only emitted lithic clasts from the substrata that alternated with both typical phreatomagmatic phases generating a wide diversity of pyroclastic density currents and fallout deposits, and typical Strombolian phases with explosive and effusive episodes. The sequences of deposits deducible from the resulting eruption sequences show substantial variations between cones, a sign of different types of eruptive behaviour, probably due to differences in the local substrata and its hydrogeological characteristics. Examples of this type of activity are the volcanoes of Santa Margarida, Croscat, Garrinada, Montsacopa, Can Tià and Cairat (Figs. 2.4 and 2.6) in the northern sector (Fig. 3.4), and the volcanoes of Puig d'Adri, Puig de Banya del Boc, Clot de l'Omera, Granollers and Sant Dalmai in the southern

Fig. 2.3 Landscape picture of the Croscat and Santa Margarida volcanoes, at the central part of the northern sector of La Garrotxa Volcanic Field (*Credit* Eduard Masdeu)

sector (Fig. 2.7). Table 2.1 summarises the eruptive sequences deduced for the volcanoes under study.

The variety present in the eruption sequences characterising the eruptive activity of this monogenetic volcanic field (Table 2.1) contrasts with the compositional monotony of its magmas (alkali basalts and basanites). The physicochemical characteristics of the magmas from most of the volcanoes in La Garrotxa Volcanic Field are very similar. They mainly correspond to leucite basanites, nepheline basanites and alkali olivine basalts, with a phenocryst (olivine, pyroxene, plagioclase) content up to 12 % and an aphiric-to-microcrystalline or microlitic groundmass. Ascent velocities of the order of 0.2 m/s have been calculated using the presence of large mantle-derived nodules and lower crust xenoliths in some of these magmas. Densities in the range 2700–2860 kg/m^3 and typical viscosities in the order of 10–10^2 Pa s have been calculated using standard methods based on crystal content and rock composition and assuming temperatures in the order of 1200–1300 °C. Despite the fact that some variations in the dynamics of the Strombolian episodes may be attributed to changes in magma flow conditions related to changes in the crystallinity and vesicularity (gas content) of the erupting magma, it is obvious that variations in magma physics are not the main factor behind the large diversity of eruption sequences recorded in this volcanic field.

In fact, as can be deduced from the sequences of deposits observed in each volcano, the differences in eruptive behaviour are related to the occasional interaction of the ascending magma with groundwater. Magma/water interaction is the main reason why so many of these volcanoes significantly deviate from the typical Hawaiian-Strombolian behaviour that characterises some of area's volcanoes and monogenetic basaltic volcanoes in general. The timing of the magma/water interaction and the way in which it occurred during the course of the eruptions differs considerably from one volcano to another. This contrasts with other monogenetic volcanic fields where eruptions seem to follow more constant patterns. In the present case, however, the large diversity of eruption sequences observed can be explained by variations in the stratigraphy and structure of the shallow geological basement beneath each volcano and the hydraulic characteristics of each aquifer.

Two main sedimentary aquifers have been identified that may have interacted with the magmas in La Garrotxa and thus given rise to a wide variety of phreatomagmatic episodes and eruption sequences. One aquifer is located at an average depth of a few hundred meters below the volcanic cones, while the other is much shallower, just a few dozen metres below the surface. The deep aquifer corresponds to Eocene continental sediments (the Bellmunt Formation), composed of conglomerates, feldspar-rich sandstones and red mudstones, while the shallow aquifer corresponds to Quaternary volcanic and alluvial deposits, mostly consisting of unconsolidated gravels, sands and clays, and volcanic products (lavas and pyroclasts) from former eruptions. A third aquifer played a significant role in some of the most important eruptions (Puig de Banya del Boc and Crosa de

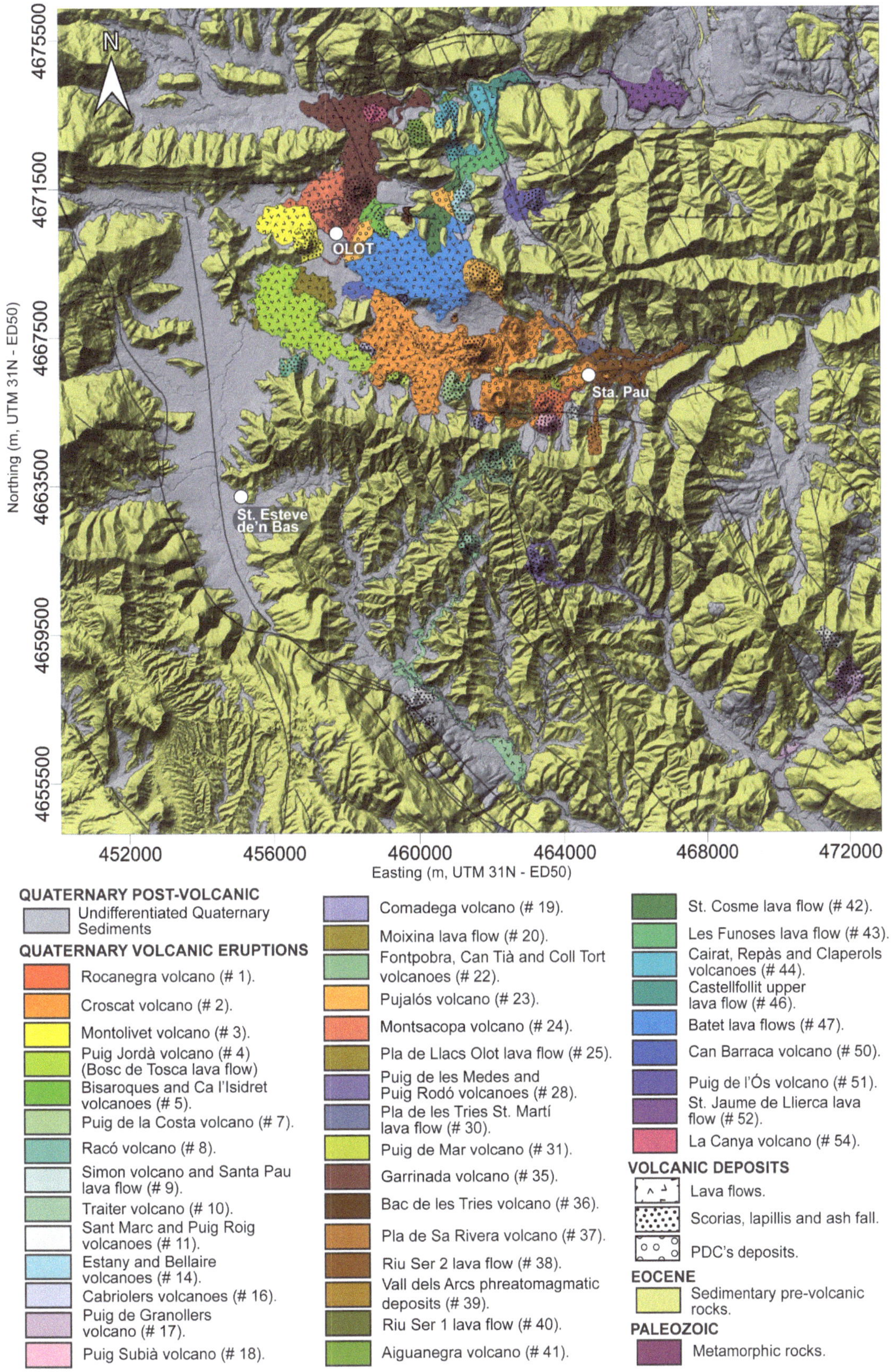

Fig. 2.4 Volcanic stratigraphy of the northern sector of La Garrotxa Volcanic Field, where all stratigraphic members (products of each single eruption) are identified with a different colours (reproduced with permission of John Wiley and Sons, Ltd.)

Fig. 2.5 Field pictures of some strombolian cones from the northern sector of La Garrotxa Volcanic Field (*Credit* Eduard Masdeu)

Sant Dalmai) and corresponds to highly fractured Palaeozoic rocks. The depth of this latter aquifer depends on the tectonics of each area and it may be a few hundred metres deep or lie at somewhat shallower depths.

In summary, the volcanoes of La Garrotxa Volcanic Field offer the opportunity to conduct different case studies that demonstrate how complex monogenetic basaltic volcanism may ensue in even a relatively small area if erupting magmas

Fig. 2.6 Field pictures of some cones containing phreatomagmatic episodes from the northern sector of La Garrotxa Volcanic Field (*Credit* Eduard Masdeu)

interact with the groundwater. This is particularly relevant when aquifers with different hydraulic characteristics are present, and when the structure of the terrain is complex due to local tectonics and/or differences in stratigraphy. The large diversity of eruption sequences deduced from the volcanoes in La Garrotxa reveal that most of the variables that have controlled them depend on local geology rather than on the magma, which can be considered as constant.

Fig. 2.7 Aerial views of La Crosa de Sant Dalmai and Puig de la Banya del Boc phreatomagmatic edifices from the southern sector of La Garrotxa Volcanic Field (*Credit* Eduard Masdeu)

Table 2.1 Summary of the different eruptive sequences deduced for the studied volcanoes

Volcano type	**Eruptive sequence** (from beginning to the end)
Puig d'Adri	Phreatomagmatic-Strombolian-Phreatomagmatic-Strombolian-Hawaiian
Crosa de Sant Dalmai	Phreatomagmatic-Strombolian-Phreatomagmatic-Strombolian
Santa Margarida-Croscat	Phreatomagmatic-Strombolian-Phreatomagmatic-Hawaiian
Garrinada	Hawaiian-Strombolian-Phreatomagmatic-Hawaiian
Can Tia	Phreatomagmatic-Strombolian-Phreatomagmatic
Montsacopa	Hawaiian-Strombolian-Phreatomagmatic
Puig de Banya de Boc	Phreatomagmatic-Strombolian-Hawaiian
Cairat	Phreatomagmatic-Strombolian
Clot de l'Omera	Phreatomagmatic
Roca Negra	Strombolian

2.3.2 Stratigraphy: The Volcanic Deposits

Volcanic terrains are characterised by the complex stratigraphic relationships of their products caused by, variously, extremely rapid depositional rates compared with conventional sub-aerial and submarine sedimentary environments, the friable character of many volcanic deposits that facilitates their rapid erosion, the contrasting influence of topography on the emplacement of volcanic products, and rapid changes in facies. Such complexity constitutes one of the main obstacles when attempting to reconstruct the geological evolution of volcanic areas. This is even more problematical in highly urbanised areas where construction modifies the original morphology of the area and hides many natural features, and where deposits are often lost to quarrying.

In the case of La Garrotxa Volcanic Field, an important part of the northern zone now lies beneath the city of Olot (almost 40,000 inhabitants), a highly industrialised and built-up area covering about 30 km^2. Deposits from five recent volcanic cones (Montolivet, Montsacopa, La Garrinada, Bisaroques and Ca l'Isidret) located in the heart of the city (Fig. 2.8) have been partially covered by or removed due to urban and industrial construction. Furthermore, a dense carpet of vegetation covers other parts of this volcanic field and thus the surface geology of the area is not always visible. Therefore, in addition to extensive fieldwork, a study of ephemeral outcrops and the stratigraphic logging of new wells and geotechnical drill holes, complemented by existing information gathered by recent geophysical studies of the substrata of this volcanic field, are required if we are to generate a comprehensive volcanic stratigraphy of the area that can identify the products of each eruption, their relative stratigraphy and their surface area.

The digital volcano-stratigraphic map of La Garrotxa Volcanic Field (1:2000) summarises all the stratigraphic members (i.e. all deposits from the same eruption) that can be identified in the region. Members are placed according to their relative stratigraphy from the youngest (top) to the oldest (bottom) (Fig. 2.4). Also given is an indication of whether the age is relative (i.e. established on the basis of observable stratigraphic relationships) or absolute (i.e. a geochronological—either isotopic or palaeontologic—date is available), as well as the corresponding reference source and the calculated erupted volumes (in Dense Rock Equivalent, DRE) when available.

The precision required to establish the stratigraphic relationships between the volcanic materials of the area decreases progressively with their age, as the younger materials are better exposed and preserved, and are located at shallower depths than older ones, often only accessible via boreholes and wells or at deep ephemeral outcrops. Due to incomplete exposure, some of the older units could have been overlooked.

The volcanic materials consist of lava flows and different types of fallout and pyroclastic density current deposits derived from Strombolian and phreatomagmatic eruptions. The calculated volumes shown in Table 2.2 are all minimum estimates of the DRE and represent the minimum volume of magma erupted in each case and, consequently, the magnitude of the eruption. Not all calculated volumes have the same degree of accuracy, since not all deposits are completely exposed. This implies that the record presented here is incomplete and should be treated with care if used in scientific studies. The volumes obtained lie in the range 0.01–0.03 km^3. However, a few eruptions (Croscat (2), Puig Subià (18), Crosa de Sant Dalmai (21), Puig de Banya del Boc (32), Garrinada (35), Puig d'Adri (45) and Puig de l'Os (51)) generated larger volumes, for example up to 0.2 km^3 in the case of Croscat.

Rather than being attributable to major tectonic events, the observed stratigraphic discontinuities and unconformities can be ascribed to either minor, inter-eruptive and syn-eruptive erosional and depositional episodes, to non-depositional hiatus corresponding to the non-uniform aerial distribution of the volcanic materials erupting from the vents, or—above all—to the fact that volcanic materials do not always emplace according to the principle of superposition, in particular when younger materials emplace on deeply eroded older ones. The style of the volcanism shows no significant variation along the stratigraphic succession and so the differences observed in the deduced eruption sequences (e.g. the alternation of Strombolian and phreatomagmatic phases) of the different members should be interpreted as due to local variations in the characteristics of the substrata rather than to any geological changes occurring on a major scale.

2.3.3 Structural Controls of the La Garrotxa Volcanic Field

La Garrotxa Volcanic Field is mostly controlled by two major Neogene faults, the Amer and Llorà faults. Both are oriented NW-SE, as are most of the major post-Alpine extensional faults that have defined a horst and graben structural pattern in NE Iberia. However, most of the eruptive fissures and secondary structural lineaments that control the precise location of the eruption centres of this volcanic field show a NNW-SSE trend that runs slightly obliquely to the main faults.

The pattern shown by the eruptive fissures and subordinated structural lineaments is compatible with a slight

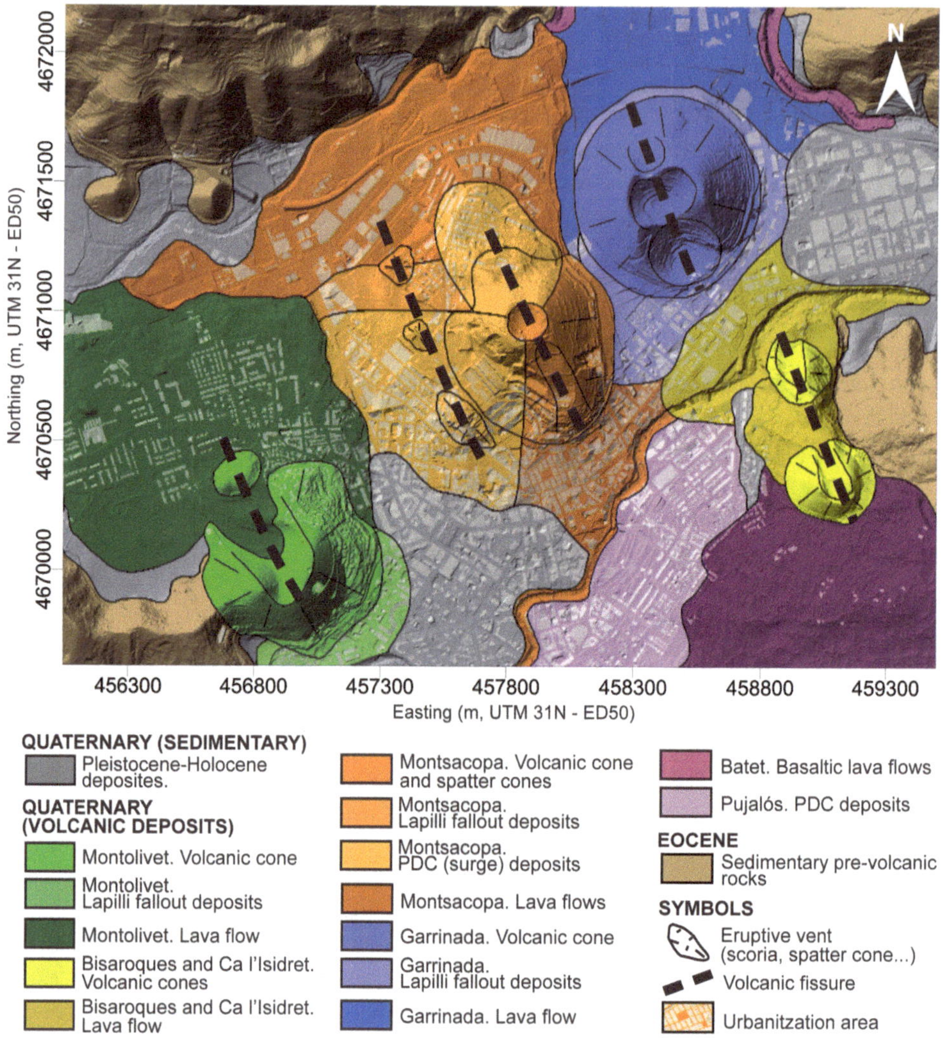

Fig. 2.8 Detailed geological map of Olot city (reproduced with permission of John Wiley and Sons, Ltd.)

dextral trans-tensional component in the two main faults and indicates that magma ascent in the uppermost crust and subsequent eruptions were controlled by these subordinated fissures. This structural configuration would have favoured the opening of these fractures and the transport of magma through them, at least in the final pre-eruptive stages. Similar behaviour has been described from other volcanic zones around the world. However, it is not clear how deep these fractures are nor what exactly their role was in transporting magma from deeper levels.

The presence of mantle-derived xenoliths and lower-crust cumulates suggests that the magmas that erupted in La Garrotxa Volcanic Field came either directly from the source region in the upper mantle or from intermediate reservoirs located at the base of the crust. No evidence exists for the presence of shallower reservoirs beneath this volcanic field. This is also confirmed by existing petrological and geochemical data that show the primitive or relatively poorly evolved character of these magmas. Therefore, what remains to be discovered is how deep the structures observed on the surface extend and how they were used by magma to reach the surface. It is worth noting that most of the area's freshwater springs are associated with relatively shallow stratigraphic or structural discontinuities, while the presence of high 222 Rn and CO_2 concentrations mainly occurs along the major Neogene faults (Fig. 2.9), thereby indicating that the deep circulation of these mantle-derived gases is permitted through these faults. Likewise, recorded seismicity also reveals that Neogene faults have been active recently, mainly linked to the Amer fault. Finally, the obtained magma ascent rates indicate that only a relatively short time was required for magmas to reach the surface, thereby suggesting the existence of preferential pathways for magma when ascending from source or accumulation regions up to the eruption sites.

In light of this evidence, the Llorà fault can be judged to have played the most important role in conducting magma from the source region, either to the base of the crust where magma occasionally accumulated (underplating) and differentiated, or directly to much shallower levels where it was

Table 2.2 Volcanic stratigraphy and volume estimates of La Garrotxa Volcanic Field

Time period	Relative Stratigraphy (from younger to older)	Age	Member	Description	DRE (Km^3)	Source
Holocene	1	Relative	Rocanegra volcano	These deposits overlie the Croscat volcano (# 2) phreatomagmatic deposits and are not covered by the Croscat volcano fallout.	0,01	(1)
Upper Pleistocene	2	11.5–13 ka	Santa Margarida – Croscat – Pomareda eruptive fissure		0,18	(2) (3)
	3	Relative < 18000	Montolivet volcano	These deposits overlie those of Montsacopa volcano (# 24), which in turn overlay La Garrinada (# 35) volcano deposits.	0,02	(4)
	4	17.1 ka	Puig Jordà volcano	Lava flow with rootless volcanoes, locally known as Bosc de Tosca.	0,03	(2) (3)
	5	Relative	Bisaroques and Ca l'Isidret volcanoes	Young aspect and poorly eroded materials overlying La Garrinada (# 35) volcano deposits.	0,01	
	6	Relative	Puig de Martinyà volcano	There is a paleosoil that separates these deposits from the Croscat volcano (# 2) deposits.		
	7	Relative	Puig de la Costa volcano (Lava flow below Croscat volcano member (# 2))	Several lava flows were recognised in the Croscat borehole below the Croscat and Puig Martinyà (# 6) deposits. The uppermost lava flow can be stratigraphically correlated with the Puig de la Costa volcano, so it should be younger than 30 ka.		
	8	Relative	Racó volcano		0,02	
	9	28.1 ka	Simon volcano and Santa Pau lava flow	Santa Pau lava flow could be related to the Simon volcano according to its location and stratigraphic position.		(2)
	10	30 to 46.3 ka	Traiter volcano		0,03	(2)
	11	Relative	Sant Marc and Puig Roig volcano			
	12	Relative	Lava flow below Bosc de Tosca	Appears underlying the Bosc de Tosca (Puig Jordà volcano (# 4)) lava flow and Racó volcano (# 8) deposits.		
	13	Relative	Pla de Massandell borehole upper lava flow			
	14	Relative (< 45 ka)	L'Estany and Bellaire volcanoes		0,02	(1)
	15	Relative	Pla de Massandell borehole lower lava flow			
	16	Relative	Cabriolers volcanoes			
	17	Relative	Puig de Granollers volcano		0,03	
	18	Relative	Puig Subià volcano	Its deposits appear below the Santa Pau lava flow (# 9).	0.1–0.2	
	19	Relative	Comadega volcano	Underlying deposits from Croscat and Santa Margarida (# 2) and other volcanoes deposits in the Santa Pau area.		
	20	Relative	La Moixina lava flow			
	21	43–120 ka	Crosa de Sant Dalmai volcano		0,13	(1) (5)
	22	73.5 ka	Fontpobra, Can Tià and Coll Tort volcanoes		0,03	(2)
	23	Relative	Pujalós volcano		0.03–0.07	
	24	Relative	Volcà del Montsacopa	Its deposits overlay La Garrinada volcano (# 35) and are covered by thus from the Bisaroques and Ca l'Isidret volcanoes (# 5).	0.02–0.06	
	25	Relative	Pla de Llacs lava flow			
	26	Relative	Pla d'Olot lava flow	Below Pla de Llacs (# 25). Only observable in boreholes.		
	27	Relative	Pla d'Olot lava flows	Undetermined number of lava flows that fill the Olot depression.		

(continued)

Table 2.2 (continued)

Time period	Relative Stratigraphy (from younger to older)	Age	Member	Description	DRE (Km^3)	Source
	28	100 ka	Puig de les Medes and Puig Rodó volcanoes		0,02	(2)
	29	Relative	Lava flow in borehole at Croscat northern flank			
	30	110 ka	Pla de les Tries lava flow			(2)
	31	110 ka	Puig de Mar volcano			(6) (2) (7) (6) (2) (7)
	32	121 ka	Puig de la Banya del Boch and Clot de l'Omera volcanoes			(2) (3)
	33	Relative	Olot 2 lava flow	Only observable in boreholes.		
	34	Relative	Olot 1 lava flow	Only observable in boreholes.		
Middle Pleistocene	35	133 ka	Garrinada volcano		0,08	(2) (8)
	36	Relative	Bac de les Tries volcano	Deposits observable below la Garrinada volcano (# 35).		
	37	Relative	Pla sa Ribera volcano			
	38	Relative	Riu Ser lava flow 2	Appearing above the Riu Ser lava flow 1 (# 40).		
	39	Relative	Vall dels Arcs phreatomagmatic deposits	Underlying the Riu Ser lava flow 2 (# 39).		
	40	Relative	Riu Ser lava flow 1			
	41	150 ka	Sant Joan les Fonts lava flow and Aiguanegra volcano			(8)
	42	Relative	Sant Cosme lava flow (from Batet)	Above the Funoses lava flow (# 42).		
	43	Relative	Les Funoses lava flow	Above the Castellfollit lava flows (# 46 and # 47).		
	44	Relative	Cairat-Repàs-Claperols eruptive fissure			
	45	168 ka	Puig d'Àdri volcano		0,09	(9)
	46	192 ka	Castellfollit de la Roca upper lava flow			(8)
	47	217 ka	Castellfollit de la Roca lower lava flow			(8)
	48	Relative	Batet upper lava flows			
	49	247 ka	Batet middle lava flows			(2)
	50	Relative	Barraca volcano	These products are covered by the Batet lavas (# 48 and # 49).		
	51	250 ka	Puig de l'Ós volcano		0,02	(8)
	52	Relative	Sant Jaume de Llierca lava flow			
	53	Relative	Batet lower lava flows	Lava flows identified in a borehole at the Batet high		
	54	Relative	La Canya volcano			
	55	590–700 ka	Sant Joan les Fonts lava flows	The oldest lava emission recognised in this volcanic field.		(8)

RS, relative stratigraphy (from younger to older)

Source: (1) Burjachs (1985), 14C dating. (2) Guérin and Valladas (1980), thermoluminiscence dating of volcanic plagioclases. (3) Guerin and Benhamou (1985), thermoluminiscence dating of volcanic plagioclases and K/Ar isotopic dating. (4) Bolos (1925), relative dating by biostratigraphy. (5) Pedrazzi et al. (2014), U–Th and 14C dating. (6) Mallarach (1981, 1982, 1998), relative dating by stratigraphy. (7) Donville (1973), K/Ar isotopic dating. (8) Lewis (2000), Ar/Ar isotopic dating. (9) B. Gómez (personal communication, 2014), thermoluminescence (TL) dating of sediments

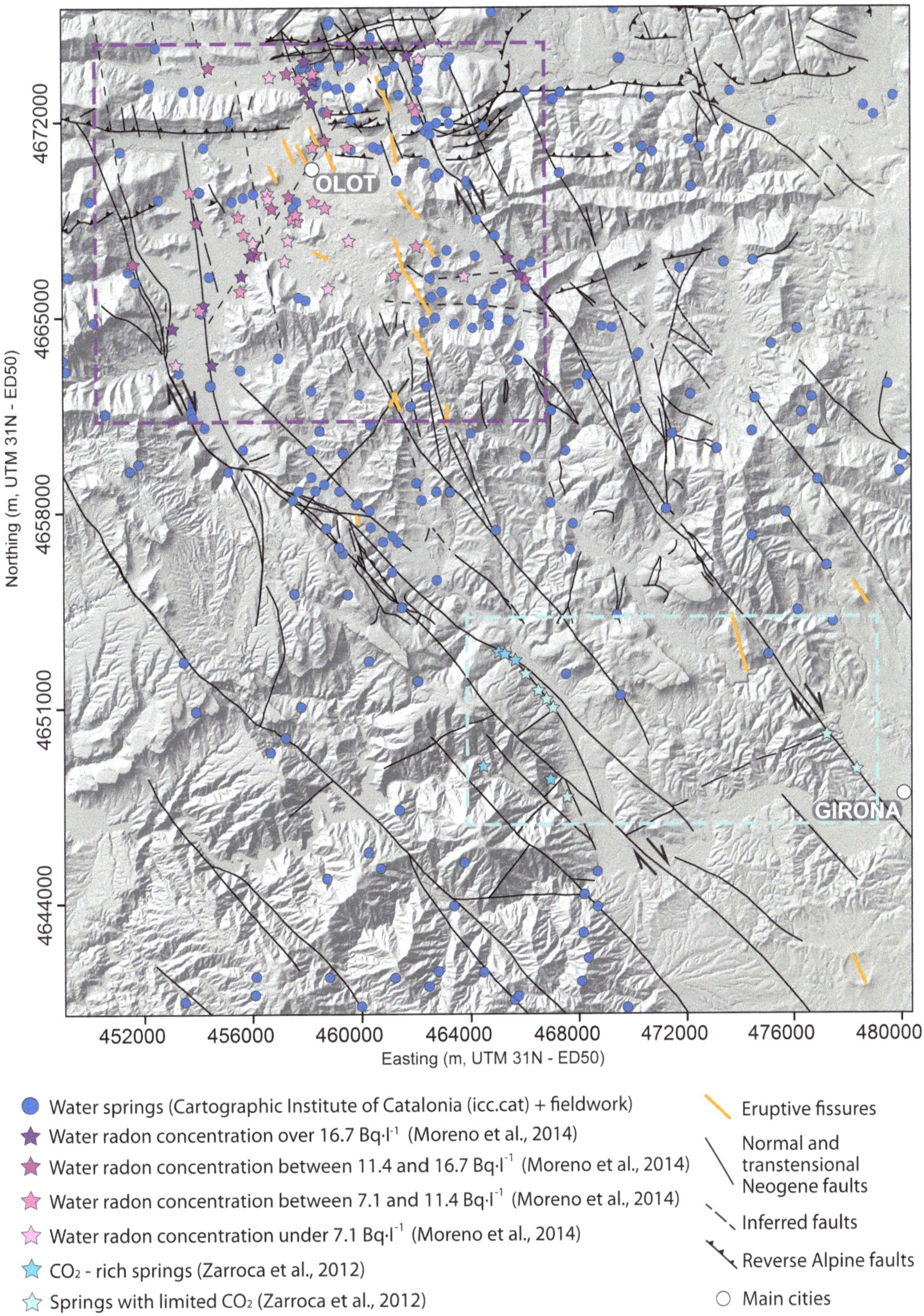

Fig. 2.9 Location of the main water springs and gas emission sites at La Garrotxa Volcanic Field (reproduced with permission of Elsevier)

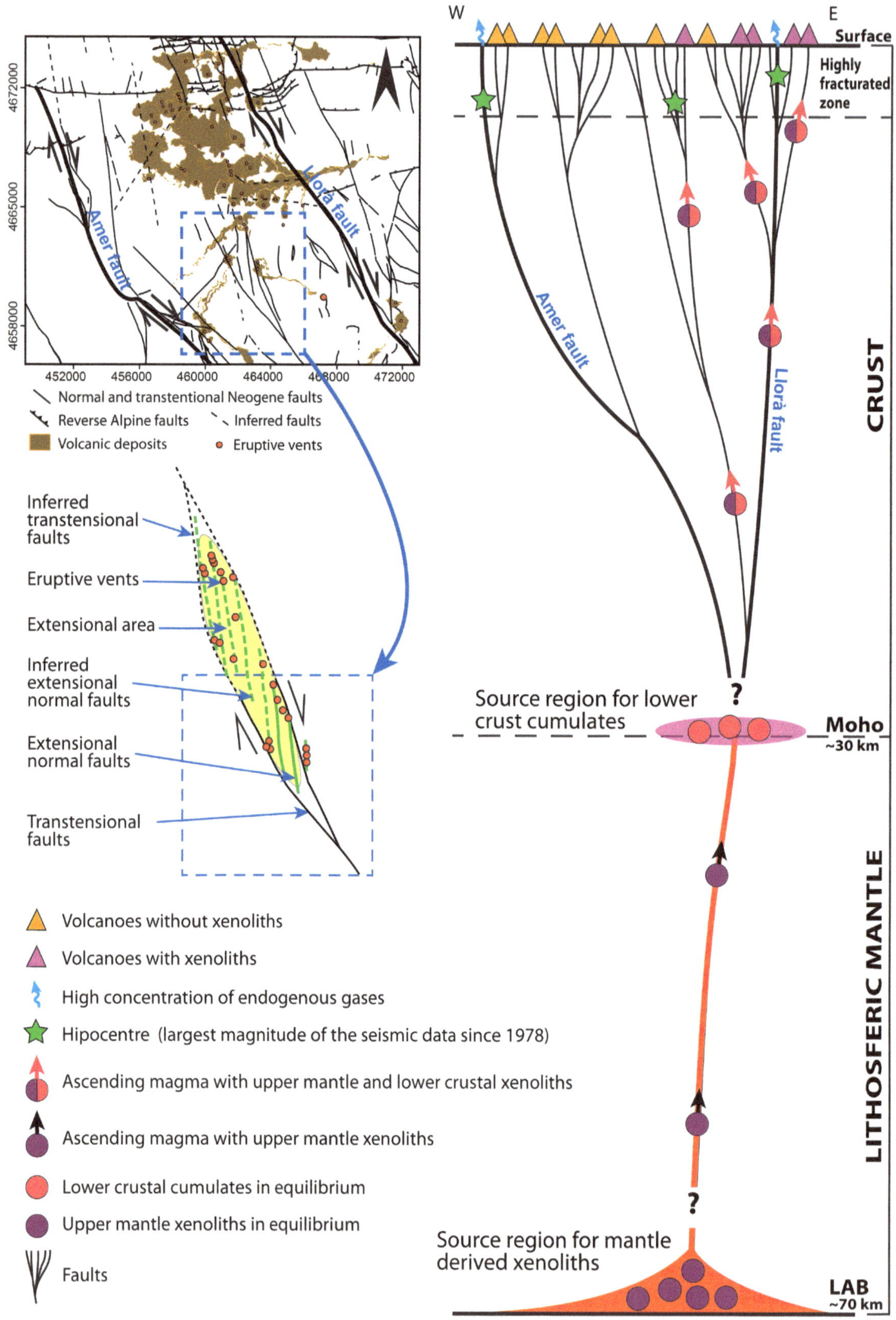

Fig. 2.10 Conceptual model of the magmatic plumbing system at La Garrotxa Volcanic Field (reproduced with permission of Elsevier)

captured by the subordinated fractures that allowed it to erupt to the surface. As indicated by the presence of lower-crust cumulates and the location of vents and eruptive fissures, the extraction of magma from the lower crust reservoir(s) was mainly controlled by the Llorà fault and, to a lesser extent, by the Amer fault, which would have acted as a conjugate major fault reaching down to the base of the crust. Shallow subordinated fractures to the Amer fault would have also captured the ascending magma in its last stages and have controlled its eruption to the surface (Fig. 2.10). The fact that the mantle-derived xenoliths and the largest lower-crust cumulates are restricted to the vents related to the Llorà fault is strong support for this model. Likewise, the volcanoes associated with this fault generated the largest erupted volumes of all volcanoes in this volcanic field, which also suggests that the Llorà fault was the main magma pathway.

The reason why magma eruptions have occurred through different NNW-SSE eruptive fissures during the history of La Garrotxa Volcanic Field—despite the constant and common feeding system at depth—seems to be related to the role played by these subordinate shallow fractures. They captured magma during the final stages of its ascent to the surface and so determined the point of each eruption. The shallow character of these fractures suggests that the local stress field, which was mostly controlled by the movement of the two main (Llorà and Amer) faults, only made a weak contribution. Under these circumstances, these shallow fractures could be easily sealed by residual magma that solidified within, thereby making it easier for a new eruptive episode to open a fresh fracture than to reuse a previously sealed one. This would also suggest—in accordance with the intermittent character of this volcanism—that each eruptive episode corresponds to an intermittent reactivation of the main fault system every 10,000–30,000 years. These tectonic reactivations would permit the ascent of deep magma and the opening of subordinate fractures in the uppermost crust, which would then erupt on the surface each time in a different location in the volcanic field.

Habitats of La Garrotxa

3

Xavier Olive

Relief, diversity of substrata, climate and the geographical position of La Garrotxa in the eastern Pyrenees, a region with a fascinating biological history, are the principal factors that explain the great diversity of species and habitats found in this county.

Nevertheless, intense anthropic disturbance has also played its part and agriculture, animal husbandry, forestry and industry have helped increase the diversity of environments and afforded greater importance to the secondary communities and open spaces present in this region.

Much of the eastern half of La Garrotxa enjoys a Mediterranean climate characterised by hot dry summers and mild winters with few frost days, and is thus dominated by typical coastal Mediterranean habitats. The principal period of activity of many of the plants in these environments is spring, between the winter and summer periods of quiescence, although autumn rains do sometimes permit a second flowering of certain species.

The commonest vegetation communities here are evergreen forests and shrublands, typically consisting of holm oak *Quercus ilex* or Aleppo pine *Pinus halepensis* formations accompanied, above all, by thermophile species that are resistant to the summer drought. In the eastern- and southernmost parts of the county, plants such as myrtle *Myrtus communis* that are more habitual in coastal areas are found.

Higher and cooler areas are dominated by sub-Mediterranean habitats, very often composed of deciduous downy oak *Quercus pubescens*—often mixed with holm oaks—and box *Buxus sempervirens* formations.

However, in the Olot basin and the mountains of the Serralada Tranversal in the west and northwest of La Garrotxa, greater rainfall and fewer sunshine hours due to cloud masses and summer storms ensure that true central European habitats, including humid deciduous woodland, appear. In the most humid areas, beech *Fagus sylvatica* forests occupy rocky, well-drained areas, while damp pedunculate oak *Quercus robur* formations, often accompanied by other deciduous trees, thrive in areas with deeper, more evolved soils. Here, relict populations of central European plants such as rue-leaved isopyrum *Isopyrum thalictroides*, spring fumewort *Corydalis solida* and common Solomon's seal *Polygonatum multiflorum*, which are as a rule rare in Catalonia, find appropriate habitat.

In the higher parts of the mountains of the Serralada Tranversal in the west of the county and in a small part of the westernmost sector of L'Alta Garrotxa, there is no summer drought and winter temperatures are relatively low; these areas harbour habitats including humid deciduous forests that are characteristic of the Pyrenean montane stage. In these environments, downy oak and beech forests, as well as mesophile grasslands populated by plant species from humid environments, occur alongside high-level scrub formations containing relict subalpine plants such as mountain pine *Pinus uncinata*, *Endressia pyrenaica* and petty whin *Genista anglica* that are here isolated from their main Pyrenean populations.

Human activity is responsible for the creation of the mosaic of vegetation types that has enhanced the territory's biological richness; nevertheless, anthropic activity has also negatively affected—in some cases severely—certain species and habitats, above all in the county's valley bottoms and wetlands. A reflection of this intense land use is the lack of diversity that occurs in some of the region's best-studied biological groups.

A number of old forests and well-managed humid hay meadows on the least accessible plateaus, as well as a few well-preserved lakes and ponds with good quality water, still provide habitat in La Garrotxa for rich communities of vascular and non-vascular plants, fungi and lichens. Many of these plants are rare elsewhere and these remaining well-preserved sites have become hotspots in Catalonia of species richness. The same occurs in sanctuaries for faunal

X. Olive (✉)
Institut Català d'Historia Natural, Olot, Spain
e-mail: xevioliver@gmail.com

J. Martí and L. Planagumà (eds.), *La Garrotxa Volcanic Field of Northeast Spain*,
Geoheritage, Geoparks and Geotourism, DOI 10.1007/978-3-319-42080-6_3

biodiversity in the case of the few animal communities—for example, birds, dragonflies and grasshoppers and crickets—for which comprehensive datasets are available. The presence of these increasingly isolated biodiversity hotspots only serve to highlight the fact that in much of La Garrotxa habitats are becoming less and less diverse due to a drop in habitat quality.

Despite these considerations, La Garrotxa is still well known for its rare species. A number of endemic taxa, found only in this county or in nearby regions, are known from the area. These include the whole world population of the milkwort *Polygala vayredae*, as well as a number of other endemic species with slightly larger ranges embracing the eastern Pyrenees and the Serralada Transversal (e.g. the plant *Lithodora oleifolia*, the cricket *Dolichopoda lindeni*, the Blind Scorpion *Belisarius xambeui* and the snail *Bofilliella subarcuata*) that have been recorded almost exclusively from the county of La Garrotxa.

Another important group are the species that, despite having wider distributions, still only have very localised populations. Aside from a few other sites in Catalonia, the records from La Garrotxa of the umbellifer *Dichoropetalum schottii*, the liverwort *Mannia fragans,* the moss *Hygroamblystegium tenax*, four species fungi including *Chionosphaera apobasidialis*, and nine of lichens including *Vezdaea dawsoniae*, are the only ones from the whole of the Iberian Peninsula.

Fig. 3.1 Coastal holm-oak forest species (viburnum, lentisc, strawberry-tree, Mediterranean honeysuckle or *Smilax aspera*). *Credit* Xavier Oliver

3.1 Mediterranean Habitats

Holm-oak forests are the typical forest type in Mediterranean zones (Fig. 3.1). This tree is well adapted to Mediterranean climates since it is able to resist summer droughts and the occasional winter frost without losing its leaves. Many of the plants in this forest formation are sclerophyllous and endure high temperatures and the lack of rainfall in summer by closing their stomata to prevent the loss of water by transpiration. They do not lose their leaves seasonally and are active in favourable environmental periods regardless of the time of year.

The tree layer is dominated by the holm oak, here generally only 10–20 m in height due to the continuous exploitation that these forests have suffered over the centuries. In places, human activity has all but eliminated this tree and in such degraded areas Aleppo *Pinus halepensis*, umbrella *Pinus pinea* and black *Pinus nigra* pines and cork oaks *Quercus suber* thrive, accompanied by a dry, fairly open understorey replete with woody Mediterranean shrub and scrub species (Fig. 3.1).

3.1.1 Coastal Holm-Oak Forests

In the most Mediterranean areas in the northeast and southeast of La Garrotxa, as well as in neighbouring areas of the counties of La Selva and El Gironès, holm-oak forests are frequent on all types of substrata between 100 and 900 m. These coastal holm-oak forests are dense, shady and impenetrable, and are reminiscent of subtropical formations due to the dominance of generally evergreen sclerophyllous plant species whose coriaceous leaves are well adapted to the prevailing environmental conditions. A few broad-leaved laurifolius species with flexible stems, hooks and tendrils that enable them to spread freely like lianas are also common. The great diversity of species is especially apparent in the shrub layer, where species such as viburnum *Viburnum tinus*, Mediterranean buckthorn *Rhamnus alternus*, *Phillyrea media*, lentisc *Pistacia lentiscus* and strawberry-tree *Arbutus unedo*, and lianas including Mediterranean honeysuckle *Lonicera implexa*, evergreen rose *Rosa sempervirens* and *Smilax aspera*, all abound.

3.1.2 Continental Holm-Oak Forests

In cooler areas, normally at higher altitudes, continental holm-oak forests that differ greatly from their coastal counterparts dominate. Inside these formations, there is no dense layer of shrubs or lianas as the climate is too cool for the thermophile species that thrive in the coastal holm-oak forests. Thus, upland species of plant that are more typical of the sub-Mediterranean downy oak forests appear. In La Garrotxa, this type of holm-oak forest covers large areas of territory lying between the warmer lowland areas with coastal holm-oak forests and the sub-Mediterranean forests that develop in dry but colder environments.

3.1.3 Mediterranean Shrub and Pine Forests

Over the centuries, large areas of holm-oak forests have been exploited to provide firewood or charcoal, or cleared for planting crops. However, subsequently, once this cleared land was abandoned in rural areas, shrubs encroached and today woody formations are often found beneath stands of 10–20-m tall pines (Fig. 3.2). The species that dominate these Mediterranean formations vary but are usually low (0.5–3-m high) and grow in fairly open habitats (depending on the human use made of them and the depth of the soil) that harbour a great variety of shrub species.

Basic soils—treeless or with stands of Aleppo or black pines—develop either impenetrable garrigues of holly oak *Quercus coccifera* or scrub dominated by thyme *Thymus vulgaris*, rosemary *Rosmarinus officinalis*, broad-leaved lavender *Lavandula latifolia*, winter heath *Erica multiflora*,

Fig. 3.2 Pine woodland with shrub formations. *Credit* Xavier Oliver

Genista scorpius, lentisc *Pistacia lentiscus* and *Globularia alypum*.

On leached sandy soils, either treeless or with stands of umbrella pines or cork oaks, the main shrub species differ. Here, formations dominated by tree *Erica arborea* and besom *E. scoparia* heaths, sage-leaved rock-rose *Cistus salviifolius*, French lavender *Lavandula stoechas*, ling *Calluna vulgaris* and *Dorycnium pentaphyllum* all thrive.

3.1.4 Mediterranean Grasslands

Aside from shrub and pine formations, dry grasslands appear on stonier soils in Mediterranean areas where holm-oak forests would potentially grow. These grasslands are only green in wet seasons, above all in spring when plants complete their life cycles; subsequently, they dry up and take on a more grey-green hue.

In some places, dense grasslands of the grass *Hyparrhenia hirta*, 1–1.5-m high and reminiscent of savannahs and dry prairies, are dominant. Elsewhere, lower grasslands of *Brachypodium retusum* thrive, while in more upland areas grassland diversity is enriched by montane species and are thus more varied, being dominated typically by blue aphyllanthes *Aphyllanthes monspeliensis*.

3.2 Sub-mediterranean Habitats

In cooler areas where the summer drought is less patent than in the lowlands and winter temperatures are much lower, deciduous forests, whose leaves fall in autumn and do not reappear until the following spring, are dominant. A winter period of quiescence is obligatory and during this season in these practically leafless forests only a few box and holly *Ilex aquifolium* plants in the understorey offer any shelter.

3.2.1 Sub-mediterranean Oak Forests

Large areas of La Garrotxa are subject to a climate that is transitional between Mediterranean and upland Atlantic zones where, due to the still manifest effects of warm summer temperatures and a degree of summer drought, sub-Mediterranean forests form (Fig. 3.3). In these forests downy oaks dominate the tree layer, often accompanied by a

Fig. 3.3 Sub-mediterranean oakwoods. *Credit* Xavier Òliver

few holm oaks, Italian maple *Acer opalus* and whitebeam *Sorbus aria*. In the shrub layer, which is not as dense or rich as that of the coastal holm-oak forests, box (often in abundance), dogwood *Cornus sanguinea*, snowy mespilus *Amelanchier ovalis*, scorpion senna *Coronilla emerus* and common juniper *Juniperus communis* all appear. Both box and juniper are evergreen shrubs and provide refuge for many species that would otherwise not find shelter in these formations during the inhospitable winter.

3.2.2 Scots Pine Forests

Many of the downy oak forests in La Garrotxa have been cut down for timber or to create pastures. The disappearance of these forests from much of the county has, on the other hand, favoured the spread of Scots pine *Pinus sylvestris*, sometimes planted but also sometimes dispersing naturally into open areas. These pine forests are fairly open and have relatively few accompanying trees and shrubs. Although the light reaches the ground easily, there is no woody understorey that can keep temperatures and humidity relatively high, as is the case of other types of forest. The open character of these pine formations is often maintained by grazing cattle.

As animal husbandry and the exploitation of these Scots pine forests decline, the typical plants that accompany the downy oak forests begin to reappear. The slow-growing pines are gradually shaded out by the relentless growth of the often initially very dense understorey, which eventually comes to be dominated by the shrubs typically found in the deciduous oak forests.

3.2.3 Blue Aphyllanthes Grassland

Open areas in the downy oak and pine forests are usually occupied by blue aphyllanthes formations, typically on dry stony soils on south-facing slopes or on deeper, humid clayey soils on north-facing slopes. These formations are accompanied by many herbaceous plants and low shrubs such as *Genista scorpius* and thyme, which flower in spring but then wither in summer.

3.3 Central European Environments

In La Garrotxa the deciduous forests that most resemble their central European counterparts appear on north-facing slopes and in valley bottoms. They are present above all in the Olot basin and in the mountains of the Serralada Transversal, two areas of greater rainfall and, more importantly, no summer drought given that the summer rainfall, fogs and clouds here substantially reduce the loss of water to transpiration.

The mosaic of deciduous woodland, thickets, shrublands, humid grassland and fields in these environments hosts a wide diversity of plant and animal species that are often much commoner on the northern face of the Pyrenees and in central Europe. Aside from the deciduous trees that form the forests here, also present are species such as rue-leaved isopyrum and the Willow Tit *Parus palustris*, a bird of humid woodland, that are normally very rare so far south of the Pyrenees.

One of the most obvious ways in which the fauna of these central European forests differs from that of the county's holm-oak forests is the lack of more thermophile avian species such as Dartford *Sylvia undata* and Sardinian *S. melanocephala* Warblers and the Woodchat Shrike *Lanius senator*, which are replaced by Bullfinch *Pyrrhula pyrrhula*, Chiffchaff *Phylloscopus collybita*, Greater Spotted *Dendrocopos major* and Lesser Spotted *D. minor* Woodpeckers, and Nuthatch *Sitta europaea*.

3.3.1 The Beech Forests

In terms of surface area, the most important deciduous forests in La Garrotxa are the beech forests (Fig. 3.4). This tree requires humid environments with well-drained soils, conditions that it finds on steep, fairly sheltered north-facing slopes in many places in the county, above all in the mountains of the Serralada Transversal that are wetter and cloudier in both spring and summer.

These forests have a dense tree layer, consisting practically only of beech with just the occasional small-leaved lime *Tilia cordata* or wild cherry *Prunus avium*. The beech branches form dense canopies whose leaves, once they appear in spring, only allow a little light to reach the forest floor, thereby hampering the completion of the life cycle of understorey plants. Consequently, these plants—many of which are bulbs or spread by rhizomes—unfold their stems, leaves, flowers and fruit in early spring before the leaves appear and whilst enough light reaches the forest floor to enable them to photosynthesise efficiently.

In beech forests on deep rich soils, as in the Puigsacalm sector in the west of the county, the understorey layer of plants is generous and forms an exuberant carpet of leaves and flowers in spring, which disappear in summer leaving behind just the underground bulbs and rhizomes whose nutrient reserves will allow them to sprout again the following spring.

On the other hand, if the soil is poor or stony as in the famous D'en Jordà beech forest that grows on a lava flow,

Fig. 3.4 Beech forests and lowland humid forests. *Credit* Xavier Oliver

there is generally only a very impoverished understorey under the dense tree canopy that is often very different in appearance from that of the rest of the county's forests. The monotony is only broken by the box bushes, which in some cases grow relatively abundantly, and the occasional holly.

3.3.2 Mixed Humid Forests with Pedunculate Oak

On rich but poorly drained soils, above all in the Olot basin and surrounding areas, pedunculate oak woodland appears, often as part of mixed humid forests accompanied by trees such as field maple *Acer campestre*, small-leaved lime and large-leaved ash *Fraxinus excelsior*.

This type of forest (Fig. 3.5) is today very rare as it potentially grows in areas that are occupied by agricultural and built-up land; thus, just a few stands of this forest survive, scattered throughout the county.

3.3.3 Riparian Woodland

Other types of humid woodland develop in La Garrotxa along streams and rivers as gallery forests. Hazel *Corylus avellana* woodland grows in galleries alongside irregular watercourses, while rivers with more or less stable flows in upland areas are lined by alders *Alnus glutinosa* or, in more lowland areas, by poplars *Populus* sp.pl. and willows *Salix* sp.pl.

Broad rivers with variable flow rates are normally lined by olive *Salix elaeagnos* and purple *Salix purpurea* willows, either alone or within the belts of alder and poplar woodland if they still exist. In the more Mediterranean areas along the river Fluvià, where the flow rate is more constant, trees such as white poplar *Populus alba* and white willow *Salix alba* are common.

In the easternmost part of the county some riparian environments host a few tamarisks *Tamarix canariensis* and *T. gallica*, as well as small stands of field elm *Ulmus minor* that are recovering after being decimated by the disease that hit them the end of the twentieth century and having been cut to create riverside cropland and pastures. These elmwoods are often associated with stands of osyris *Osyris alba* that constitute the outermost belt of the riparian woodland.

In the part of the county with the most Atlantic-type climate, this outer ribbon of elms is replaced by humid woodland composed of large-leaved ash, small-leaved lime, field maple and pedunculate oak, together with thickets of brambles *Rubus ulmifolius* and other shrubs.

3.3.4 Shrublands and Thickets

In central European environments, clearings, woodland edges and pastures tend to evolve towards shrublands composed above all of broom *Sarothamnus scoparius* and bramble, sloe *Prunus spinosa* and hawthorn *Crataegus monogyna* thickets. Although capable of invading open spaces in a very short time, these formations still allow the trees that in the future will constitute the new forest to germinate and grow.

3.3.5 Grasslands

Central European grasslands, generally forming on rich deep moist soils, are very productive and host almost exclusively herbaceous species that flower rather than wilt in summer.

If not intensively grazed, these cool humid upland grasslands are home to a notable diversity of vascular plants (on average, around 50 species), of which those that are

Fig. 3.5 On rich but poorly drained soils, above all in the Olot basin and surrounding areas, pedunculate oak woodland appears. *Credit* Xavier Oliver

typical of rich grasslands (around 30) account for over 75 % of the ground cover. Such plants include both yellow *Trisetum flavescens* and false *Arrhenatherum elatius* oat-grasses, Yorkshire fog *Holcus lanatus* and *Phleum nodosum*, and flowers such as ragged-Robin *Lychnis flos-cuculi*.

These florally diverse grasslands are also home to many faunal groups including many Orthoptera (crickets and grasshoppers). On average, 18 species of this group of insects live in these upland grasslands, an exceptionally high figure compared to the rest of the county where the average number of species in grassland varies between six and 13, and is four or less in the most artificial and pastured environments. Typical cricket species (with long antennae) in these grasslands include Grey Bush-cricket *Platycleis albopunctata*, *P. tessellata*, *Ephippiger ephippiger*, the Great Green Bush-cricket *Tettigonia viridissima* and the large *Decticus albifrons*. The most abundant grasshoppers (with short antennae) are *Calliptamus barbarus* with red wings, the Short-horned Grasshopper *Pezotettix giornae* with a pale lateral band on its thorax, and the Awl-shaped Pygmy Grasshopper *Tetrix subulata* and *T. nutans*, with the dorsal part of the thorax enlarged to cover the abdomen.

3.4 Rocky Environments

Rocky habitats—above all limestone and sandstone cliffs—are well represented in La Garrotxa.

In the southern and eastern part of the county, the sunny faces of calcareous rocks host thermophile species that can resist high temperatures and a lack of water. They generally grow as small plants with small pale-coloured leaves whose pilosity ensures that they transpire less when temperatures are high. The most abundant of this type of plants are *Sarcocapnos enneaphylla* and *Phagnalon sordidum*, whilst other rarer species found at just a few sites include *Jasonia glutinosa*, the fern *Asplenium petrarchae* and *Pimpinella tragium*.

More upland cliffs, cooler due to their greater altitude and/or north-facing positions, have a greater density and diversity of plants, which grow in the cracks and on the

Fig. 3.6 Landscape with cliffs. *Credit* Xavier Oliver

small ledges that abound in these habitats (Fig. 3.6). Of great interest are the populations of ramonda *Ramonda myconi*, the ferns *Asplenium fontanum* and Maidenhair spleenwort *A. trichomanes*, the large Pyrenean bellflower *Campanula speciosa*, Pyrenean honeysuckle *Lonicera pyrenaica* and *Potentilla caulescens*.

Siliceous rocks are essentially only present in the volcanic zone, where basalt cliffs and other volcanic materials outcrop. They are generally fairly poor in vascular plants—although it is worth highlighting the presence of creeping snapdragon *Antirrhinum asarina*, the ferns forked spleenwort *Asplenium septentrionale*, Maidenhair spleenwort *A. trichomanes* and common polypody *Polypodium vulgare*, and *Sedum hirsutum*—but are rich in lichens and mosses such as *Hypnum cupressiforme* and *Hedwingia ciliata*.

Stone walls abound in the volcanic zone and their cracks, filled with a thin layer of organic material, offer habitat for pellitory-of-the-wall *Parietaria officinalis*, rusty-back fern *Ceterach officinarum*, wall-rue *Asplenium ruta-muraria* and ivy-leaved toadflax *Cymbalaria muralis*.

3.5 Wetland Habitats

Natural wetland habitats are not common in La Garrotxa and have either been drained to create agricultural or urban land, or have been converted into simple water reservoirs. Aside from the wetland environments associated with the river Fluvià, only a few vestigial ponds and lakes of the many former much larger wetlands have survived pollution and drainage.

In the flood plains—increasingly smaller due to the canalization of the rivers—a few small temporary pools still remain alongside the main watercourses. The most stable vegetation communities appear in the river itself (Fig. 3.7), often where lakes have formed behind weirs built to divert water into irrigation channels.

Fig. 3.7 The most stable vegetation communities appear in the river itself, often where lakes have formed behind weirs built to divert water into irrigation channels. *Credit* Xavier Oliver

In the outermost part of the wetlands damp meadows and stands of round-headed club-rush *Scirpus holoschoenus* and European meadow rush *Juncus inflexus* form. Where the water is not too deep, a belt of tall bulrushes *Typha latifolia* appears, occasionally accompanied by common reed *Phragmites australis*. In the deeper centre of the wetland, hydrophyte communities thrive and consist of, above all, aquatic plants such as longleaf pondweed *Potamogeton nodosus*, amphibious bistort *Polygonum amphibium* and European water-plantain *Alisma plantago-aquatica*, whose leaves float on and completely cover the water surface.

These wetlands hold bird communities typically composed of species such as Moorhen *Gallinula chloropus*, Cetti's Warbler *Cettia cetti* and Mallard *Anas platyrhynchos*; if the wetland is larger, Little Grebe *Tachybaptus ruficollis*, Zitting Cisticola *Cisticola juncidis*, Greater Cormorant *Phalacrocorax carbo*, Grey Heron *Ardea cinerea* and Kingfisher *Alcedo atthis* may also be present.

The county also boasts one of Catalonia's most diverse and abundant populations of damselflies and dragonflies (27 species), present in three well-conserved wetlands that are relatively large and possess a wide range of micro-habitats. Typical species include the large Blue Emperor *Anax imperator* and Migrant Hawker *Aeshna mixta*, as well as the much smaller Azure Bluet *Coenagrion puella*, Blue-eye *Erythromma lindenii*, Iberian Bluetail *Ischnura graellsii*, Western Willow Spreadwing *Lestes viridis*, White Featherleg *Platycnemis latipes* and Large Red Damsel *Pyrrhosoma nymphula*. In the remaining, less well-preserved wetlands, on average just 6.7 species—most of which are generalist—are present.

History and Culture of La Garrotxa

4

Mireia Tresserras and Eva Duran

Like the rest of Catalonia, La Garrotxa is a land through which a multitude of different peoples have passed and left their mark. History knows no pauses and so should be seen as a continuum. Yet, it is often easier to study past times if they are divided into periods or 'eras'. The criteria used to establish such divisions are often highly debatable and a more realistic approach will depend on specific territorial and cultural considerations, which will either temper or reinforce any attempt at periodisation.

4.1 Prehistory

The climate of the Palaeolithic period (3,000,000–10,000,000 years ago) was characterised by a succession of ice ages and inter-glacial periods. The environment determined the nature of human activity, which in turn only had a limited effect on the natural world. Human survival was based largely on hunting and subsistence (hunter-gatherer societies) and people lived in low-density population groups of 20–30 individuals. Social organisation was rudimentary (social equality but a gender-based division of labour) and essentially nomadic, i.e. settlements were established temporarily in places such river terraces and caves. Stone tools were fashioned out of quartz and flint, two commodities that were in fact not very common in La Garrotxa.

Although the volcanic activity in La Garrotxa did not hinder the settlement of human groups in the area, it is clear that it did influence how the region was occupied. In the Lower Palaeolithic, the most ancient hominids found in Europe were the *Homo erectus* found at Atapuerca (1,500,000 years ago) and Taltaüll (700,000 years ago), who originated from Africa and established themselves in caves alongside rivers and in the open air.

The oldest evidence of human presence in La Garrotxa dates from this period and consists of remains found at SANTA HELENA (La Barroca). In the Middle Palaeolithic (80,000–40,000 years ago) *Homo sapiens neanderthalensis* hunter-gatherers journeyed along the valley of the river Llierca and, judging by the finds made in caves such as COVA 120 or COVA DELS ERMITONS, used it as a place for refuge.

The Upper Palaeolithic is characterised by the appearance of *Homo sapiens sapiens*, whose remains have been found at LES MULLERES, an encampment on the Olot plain near the town of La Canya. The archaeological material found here consists exclusively of stone tools that appear on the surface of fields. Also of interest is ROC DE LA MELCA in Sant Aniol de Finestres, a small cave on the left bank of the river Llémena.

Dating from the Mesolithic (10,000–7000 years ago), the remains of a large seasonal settlement (8000–9000 years ago) were discovered in La Garrotxa in 1991 at a site known as LA RODONA. It is thought that the existence of marshlands in the Olot plain encouraged the temporary settlement of this site during seasonal hunting-gathering activities. The distribution of the remains has enabled archaeologists to distinguish three areas that were used for different tasks: a stone-working area, an area for domestic activities, and an area for fires (lit by fungi and oak and hazel wood). The settlers lived in huts in the open air.

The Neolithic period, characterised by a more oceanic climate, originated 7000 years ago from a nucleus of settlements in the Middle East and lasted in Western Europe until 3000 years ago. Although hunting-gathering was still significant, during this period stable productive activities based on agriculture (wheat and barley) and animal husbandry (sheep, goats, oxen and pigs) began to appear and have an incidence on the landscape. Surplus production led

M. Tresserras (✉) · E. Duran
Educ'art serveis educatius i culturals s.l, Olot, Spain
e-mail: mireia.tresserras@educart.biz

E. Duran
e-mail: eva.duran@educart.biz

J. Martí and L. Planagumà (eds.), *La Garrotxa Volcanic Field of Northeast Spain*,
Geoheritage, Geoparks and Geotourism, DOI 10.1007/978-3-319-42080-6_4

to skill specialisation and the emergence of industry based on polished stone, textiles and pottery; soon stable settlements were established, which led to drastic transformations in people's ways of life.

La Garrotxa was occupied intensely for the first time in the Neolithic, as remains found at COVA S'ESPASA and COVA 120 indicate. An important population nucleus was established in the Llierca valley, where excavations in a number of caves and open-air sites occupied during the Neolithic have revealed the use made of these places as temporary settlements or more permanent settlements with, for example, storage silos hewn from the rock. PLANSALLOSA was the main open-air Neolithic settlement between the mountains of L'Alta Garrotxa and the plains at the confluence of the rivers Fluvià and Llierca, and was used for cereal cultivation. Other Neolithic settlements include CODELLA (Fig. 4.1), on the edge of the former lake in La Vall d'en Bas, and LA PRUNERA, situated at the top of the Sant Cosme pass. This latter site, where 68 hand-thrown pottery fragments and pebbles were found in a 15-m thick carbon-rich soil horizon, had no running water but its position near flat land indicates that some kind of agriculture was practised in the area.

Megalithic monuments such as collective burial mounds with dolmens and menhirs (e.g. in La Garrotxa, DOLMEN DEL PUIG DEL MORO in La Serra de les Medes, dating from the end of the Neolithic) were originally constructed with a chamber and a short tunnel that were completely covered by a circular earth tumulus. Inside Puig del Moro, the remains of nine people have been found with treasures consisting of pottery vases and other utensils. The menhir PEDRA DEL DIABLE in the Ser valley near Santa Pau is thought to have been erected as a marker, probably in the Neolithic or in the Bronze Age (3000 BC). However, the lack of any other archaeological finds in the area means that it is difficult to attribute an exact age to this standing stone.

4.2 The Bronze and Iron Ages

The Bronze and Iron Ages (5000–2500 BC in Europe) saw the dawn of metalworking: copper, bronze (an alloy of copper and tin) and then iron. In La Garrotxa metal articles have been found in cave burial sites in the Llierca valley in the COVA DELS ENCANTATS above Begudà and COVA DE ROCA LLADRE in the Serra del Corb.

Fig. 4.1 Codella, neolithic settlements on the edge of the former lake in La Vall d'en Bas. *Credit* Llorenç Planagumà

The Greeks and Phoenicians set out on long expeditions from their homes in the eastern Mediterranean towards the western coasts in search of metals, new markets and lands in which to settle. Both played a key role in the spread of metalworking techniques. The Phoenicians set up factories on, above all, the southern coasts of the Iberian Peninsula (1100 BC) and on the island of Ibiza; by contrast, the Greeks colonisers from the *polis* (or cities) of Rhodes and Phocaea in Asia Minor arrived somewhat later and founded Roses (eighth century BC) and Emporion (sixth century BC), respectively.

Although the origin of the Iberian culture has been greatly debated, it now seems clear that it represents a continuum from the indigenous Bronze Age people who came into contact with the Phoenicians and Greeks. This culture was characterised by groupings of peoples—Indigets, Ilergets, Laietans and Ilercavons, for example—who formed confederations of tribes (groups linked by family ties) that shared similar cultural traits. Agriculture, along with ironworking, weaving and the manufacture of pottery, was the basis of their economies. These societies were hierarchical in structure and boasted a degree of urban development consisting of settlements on hilltops encircled by defensive walls (e.g. Ullastret in the county of Baix Empordà). The position of these towns and villages on hilltops, coupled with eyewitness accounts by contemporary classical authors, seems to suggest that this was a time of unrest and conflict between the Iberian tribes. These peoples had a written language that today remains undeciphered.

Very little evidence exists of the presence of these tribes in La Garrotxa; the few known Iberian sites include CAN PEDREGÓS (Sales de Llierca), CAN BUCH (Santa Pau), where pottery shards were found, SANTA CECÍLIA (Sant Feliu de Pallerols), LA PALOMERA (Sant Aniol de Finestres), a settlement that seems to have acted as a look-out point, and the settlement of EL BOSCARRÓ (Sant Joan les Fonts) (Fig. 4.2).

Three small lead tablets from a tomb excavated in Empúries have Latin inscriptions on both faces that refer to a people called the Olossitani. Even though the exact significance of these texts is still obscure, some authors claim that the Olossitani were an Iberian tribe that lived in the area around Olot.

Remains have been found over an area of 1 ha and to a maximum depth of 60 cm in the Iberian settlement of EL BOSCARRÓ (Sant Joan les Fonts), one of the most

Fig. 4.2 Boscarró, Iberian site. *Credit* Llorenç Planagumà

important such sites in the region. It is thought to date from between the fourth and second centuries BC and important finds of local, as well as imported Greek, Punic and even North African pottery have been made there. These finds are significant because they demonstrate that this settlement traded with people from outside the region, a commercial practice that was more usual in coastal settlements than in those in the hinterland. It is also the first Iberian settlement known from La Garrotxa to be found so far inland.

Throughout the second and first centuries BC, Iberian society was gradually transformed by the power wielded by the Romans; new towns and cities were founded, which led to the abandoning of the old Iberian settlements and the integration of their inhabitants into the Roman way of life.

The arrival of the first Roman troops in Catalonia in 218 BC was closely linked to the struggle between the Rome and Carthage for the control of the western Mediterranean that was waged during the Second Punic War. It also signalled the beginning of the Romanisation and Roman occupation of the Iberian Peninsula (third century BC–fifth century AD), a process that led to the full but gradual transformation of the indigenous Iberian society into a community with a marked Latin accent. By the first century BC, the inhabitants of Catalonia were no longer Iberian but Hispano-Romans; this process represented a cultural and civilising phenomenon rather than any change in the ethnic make-up of the Peninsula's inhabitants, since there was no substitution of the population and relatively few settlers came to live in the new Roman colony of Hispania.

Urban life flourished during this period and the Roman population centres, either newly constructed or built upon existing Iberian nuclei, formed a dense, perfectly linked network covering the whole country that became the lynchpin of the Romanisation of the territory. The Romans introduced olive, walnut and vine cultivation and in doing so modified the landscape. It seems that the natural oak forests of La Garrotxa suffered and were gradually replaced by pines, other plant communities such as heather shrublands, which reached their zenith in this period, and cultivated areas. The Romans also left behind a great cultural legacy that included the use of Latin (Catalan is a Romance language derived from Latin), the practice of the Roman religion—from the third century AD onwards, Christianity replaced the Roman gods that had evolved from the Greek and other gods—and the implantation of Roman law.

Traces of Roman civilisation have been found in La Garrotxa at BESALÚ and include a number of pottery shards and a statue of a woman in a tunic (with her head and left arm missing). However, the most vivid evidence of the Roman influence on the region is the VIA ANNIA (Vall de Bianya–Sant Pau de Segúries).

The Roman conquest transformed the old tracks and natural pathways into a well-organised communication network. The Roman roads had military (troop movements), economic (trade between ports, markets and productive areas), socio-political (communication between cities and other settlements) and cultural (the spread of the languages and arts) functions, and played a vital role in creating and consolidating the new Roman social order and as an efficient tool of territorial organisation and exploitation.

The most important Roman road in Catalonia was the VIA AUGUSTA, which connected Rome to Cadiz. It was linked to a series of secondary roads—including the VIA ANNIA (Fig. 4.3)—that kept the most distant villas and settlements in touch with the Via Augusta. A number of sources provide information about these roads: toponyms, itineraries (private or public texts that were used as guides for travellers), milestones or *millariums* (waymarkers that indicated both distances and the names of the builders or restorers of the roads), and other archaeological remains.

Their actual routes were determined by relief features: the Via Augusta, for instance, followed the natural pre-coastal depression, while the secondary ways ran through depressions and plains, or up river valleys.

The VIA ANNIA linked the northeastern coast of Catalonia to La Cerdanya in the Pyrenees (IUNCARIA-BESALÚ-VALL DE BIANYA-COLL D'ARES) and its main function was to connect the agricultural areas of the valley of the river Fluvià to the mines in L'Alta Garrotxa and El Ripollès. It was a considerable feat of engineering as the steep gradients of the mountains obliged its builders to design a snake-like road that wound uphill along short straight stretches of road interspersed with tight curves, all resting on solid platforms and contention walls built without mortar. As well, they built drains, culverts, bridges, water deposits, defence towers and other fortifications to ensure the upkeep of the road.

Despite the lack of evidence that confirms its origin, the old road at CASTELLFOLLIT DE LA ROCA that winds up from the valley bottom to the plain is thought to be Roman. There is also evidence of the existence of the two Roman bridges at Castellfollit, one thought to correspond to the PONT TRENCAT (lit. 'Broken bridge') that once crossed the river Fluvià and the other to the ruined bridge that once crossed the Riera del Turonell here.

4.3 The Middle Ages

After the fall of the Roman Empire, the Visigoths established themselves in the Iberian Peninsula; they in turn were substituted by the Moors from North Africa, who settled in the Peninsula in the eighth century and brought to the Peninsula a civilisation that was much richer and more varied than the Christian civilisation of the time. No traces of

Fig. 4.3 The Roman, Via Annia. *Credit* Eva Duran. *Source* Educart

either the Visigoth or Moorish periods have been found in La Garrotxa.

The French kings reacted to the Moorish invasion by creating the countships of the Marca Hispanica (or Spanish March). The Carolingian conquest established a frontier with Moorish-held lands along a line that followed the rivers Llobregat and Cardoner as far as the natural barrier of the mountains of El Montsec. To the east and north of this line lay Old Catalonia (including La Garrotxa), whilst the lands to west and south were considered to be New Catalonia. The differences in the settlement patterns and many of the customs in these two territories are still recognisable today.

As is the case of other European nationalities, the historical origins of Catalonia as a nation go back to the eighth–tenth centuries. Thus, by the eleventh and twelfth centuries Catalonia as an entity existed and all the Catalan counts had been absorbed by the countship of Barcelona. A Catalan feudal state had formed and, once the south had been conquered, for the first time all the lands of what is modern-day Catalonia swore loyalty to a single authority. In the first half of the twelfth century, the Barcelona countship was united with the Crown of Aragon, thereby giving birth to the Catalan-Aragonese confederation. The thirteenth century was marked by the expansion of Catalonia in the Iberian Peninsula (Valencia) and through the Mediterranean (Mallorca), and was when the city of Olot was founded and the valley of the river Fluvià first became densely populated.

Forests were cut and land tilled and grazed, and the early Middle Ages (eleventh–thirteenth centuries) became a time of economic growth and expansion that peaked in the thirteenth and first third of the fourteenth centuries. There was also a great leap forward in commercial and artisan activities in what was still a largely agriculture-based society.

The lower Middle Ages (fourteenth–fifteenth centuries), on the contrary, were a time of structural crises that affected all facets of Catalan society: demographic (high mortality, famine, plague), social (urban conflicts, peasant revolts), economic (grave economic downturn) and political (civil war). This led to the outbreak of the first Catalan Peasant's Revolt (1462–1472), which coincided with the civil war between King Joan II and the Generalitat, the main Catalan government institution. This rural revolt was centred above all on La Garrotxa, where the warring peasants under the leadership of Verntallat from the Hostoles valley supported the king and fought, above all, from bases situated in the region's mountain areas (Fig. 4.4). Their objective was the

Fig. 4.4 Hostoles castle. *Credit* Elisenda Guitart. *Source* Tosca

abolition of the *mals usos*, the laws that divided feudal society into classes with legally different rights and obligations. Feudalism defined the links between the three societal classes that were based on two types of relationships: the socio-political lord-vassal bond defined the nexus between the privileged classes (the lord and his vassals), while the socio-economic lord-serf relationship defined the interaction between the lord and serf in each domain. During the eleventh–twelfth centuries socio-economic changes gradually transformed the peasants into dependent subjects of the large landowners (nobles and religious orders), just as socio-political changes led to the fracturing of the structures of public power and the usurpation by the feudal lords of part of this power in name of their counts.

A myriad of extraordinary data and documents on the events of this era exist in La Garrotxa. The monasteries helped spread learning and civilisation, above all in mountain areas where the clearing and settling of the land were especially challenging tasks. The repopulation of the region was begun by a group of Benedictine monks who crossed the Pyrenees in 859 and founded the monasteries of SANT ANIOL D'AGUJA and SANT LLORENS DEL MONT. La Garrotxa in this period formed part of the countship of Besalú. A second wave of colonisation was promoted by, above all, the great monasteries of SANTA MARIA DE BESALÚ (977) and SANTA MARIA DE RIPOLL; despite the distance, the latter institution became the feudal owners of the town of Olot in 1097 and continued the resettlement tasks begun by the monastery of Sant Aniol. The spread of feudalism continued as elsewhere: the lords leased their lands in exchange for the *rompuda* (the deforesting of a patch of land) with *aprisió*, that is, the right after 30 years of uninterrupted occupation to own a piece of land that had been deforested. This favoured the appearance of numerous small landowners and the territory soon became subdivided up into farms or *masos*.

During the fourteenth and fifteenth centuries the tendency in La Garrotxa was for the local population to group together in settlements with well defined but differing characteristics. For example, the original nucleus of the village of SANT FELIU DE PALLEROLS lay inside the *sagrera*, the area offering immunity within a radius of 30 ecclesiastical paces around the parish church to all those who took refuge there. This haven, which had been first defined in 1033 by the Assembly of the Peace of God (*Pax Dei*) and the Truce of Vic, guaranteed security and many local peasants built their homes in this *sagrera*; by the beginning of the fourteenth century this quarter of Sant Feliu had been totally built up. In LA VALL D'EN BAS the parish boundaries marked the limits of the different communities that had been subject to the rule of the Vice-counts of Bas or Cabrera since the end of the fourteenth century. Nevertheless, here the lordship of a domain did not correspond to a single feudal lord and, like in Sant Esteve d'en Bas, there were even serfs who had been freed of their feudal obligations. In almost all the parishes there was a manorial house belonging to the knight or bailiff who acted as the lord's representative. In the Middle Ages, the lands in SANTA PAU (Fig. 4.5) were ascribed to the counts of Besalú and were first colonised by monastery-builders. In the twelfth century the population of Santa Pau grew and parishes, homesteads and castles began to be built. Ponç III was the first recorded lord of the village and was a close companion of King Pere, joining him on an expedition to fight in Sicily. The king awarded Santa Pau the right to hold a weekly market on Mondays and in 1300 the village was granted a municipal charter. Its castle, erected on a hillock in the centre of the village, is a regular square building that was constructed in different phases during the thirteenth-fourteenth centuries. It is surrounded by a moat, which today has been converted into a square, La Placeta dels Balls. Although the origin of the counts' town of BESALÚ can be traced back to the ancient history of the Iberian, Celts and Romans, its medieval past is without doubt its most remarkable feature. Its castle, documented from the high Middle Ages (tenth century), was constructed on a strategically important knoll that today is also home to the remains of the canonical church of

Fig. 4.5 Santa Pau Medieval. Author. Pep Callís. *Source* Documentation Centre, Garrotxa Volcanic Zone Natural Park

Santa Maria. The importance of the countship of Besalú rose and fell with the pass of time and under Count Bernat I Tallaferro (990–1020) it reached its apogee, becoming an important centre of ecclesiastical power throughout the eleventh century.

4.3.1 Medieval Architecture

Romanesque art (Fig. 4.6) can be defined as a body of art—albeit with notable chronological and stylistic differences between regions—that was produced in Western Europe in the tenth to thirteenth centuries. Two periods can be distinguished, a first or 'Lombard' period and a second period from the twelfth century onwards in which art adapted to the new architectural tendencies that were developing. The architectural elements employed in the religious, civil and military buildings are characterised by rounded arches and barrel vaults that contrast with the pointed arches and ribbed vaults of the later Gothic period. In this way, artistic expression was expressed through the architecture. Romanesque churches tended to have one or three naves separated by columns that supported the weight of the walls, with an apse at the rear that was often decorated with Lombard bands, lesenes and dentate mouldings. The roofs were generally stone barrel vaults, and the main entrance was through a decorated doorway in the façade. Bell towers were of two types: a simple structure with a bell cote or a tall, square or circular, tower.

Romanesque art is an aesthetical representation—both monastic and aristocratic—of feudal society. The clericalisation of culture and the ascendency of faith over reason gave the church the right to establish in an authoritarian and unquestioned fashion the main guidelines and limits to cultural expression. Art was regarded as an extension of divine service and responded to the principles of authority, hierarchy and theocracy, which furnished a dehumanised, anti-natural and symbolic vision of reality. Romanesque structures and the iconographic murals that decorated their interior walls remind the viewer that we humans are mere subjects of a supreme being—God and, in his absence, the count, abbot or emperor—and, if in discord, we will be judged severely and deprived of our rights. This ideology exalted the idea that men and women were prisoners, condemned to a life of service, and that they can do nothing that is not consented to by the Lord.

Fig. 4.6 Romanesque church. *Credit* Llorenç Planagumà. *Source* Tosca

Gothic architecture can be defined as the artistic output in the western world that appeared approximately in the thirteenth to the beginning of the fourteenth centuries. Its construction techniques are characterised by pointed arches and ribbed vaults, along with buttresses and flying buttresses, pinnacles, fasciculated pillars, gables, spires and rose windows, together with figurative, vertical lineal symbolism. The Gothic nature of a building implied a new conception of space and, above all, the idea of continuity that unites the various elements of a building.

The cathedral is the most distinctive element of gothic architecture and the one in which most efforts were made to reach architectural solutions. The basic components—pointed arches and ribbed vaults—broke with the static nature of Romanesque architecture and bestowed buildings with a greater, more dynamic sense of energy and movement. The cylindrical pillars or piers crowned by capitals bear the weight of the structure, and are often interspersed with alternating clustered columns.

The concentration and distribution of the load free the building's walls of the need to act as mere supports and allow large windows to be encrusted into them. Thus, large stained-glass windows began to appear that tempered the light, thereby giving a more spiritual atmosphere to the body of the building. Space and light thus prevailed over the constructed mass. Buttresses rather than the abutments used in Romanesque structures were used to counteract the outward force of the vault walls, and buildings were often decorated with pinnacles and gargoyles (which acted as spouts for water running off the buttresses). The floor plan was typically basilical with a central nave (wider and higher) and two lateral naves. The outside of the building was built in the same style as the interior and reflects the laws that rule and sustain the whole. Rose windows, doorways, arches and statues fill the façade and remind us that the walls are no longer mere static objects; graceful lines dominate and throughout the *chiaroscuro* prevails.

4.4 Modern Age

The most important historical event to occur in the fifteenth century was the unification of the crowns of Castile and Aragon, out of which the new SPANISH MONARCHY was born. This new political entity was dominated by Castile, at that time immersed in an epoch of great economic, social and political effervescence. Catalonia, a small part of a large empire, retained all of its institutional and legal structures, which enabled it to remain apart from the political and economic fluctuations that ensued in Castile.

In 1427 and 1428, when Olot was still subject to the feudal lords of Ripoll and the Catalan throne was occupied by Alfons IV, the city and La Garrotxa were hit by a succession of devastating earthquakes. In light of the destruction, the city petitioned the king for permission to build a new city outside the jurisdiction of the monastery of Ripoll. The king acquiesced and the city centre thus moved from the Palau quarter to where today stand the Plaça Major and the church of Sant Esteve, the former on the edge of the new city but the latter at its heart.

The struggle between peasants and lords, which included the storming of some of the county's castles, continued throughout La Garrotxa during this period. These protests, which aimed to put an end to feudal oppression and the abuses committed by the lords, were only resolved with the signing of the Sentence of Guadalupe at the end of the fifteenth century.

The rural condition of the peasants marked four centuries of the history of Catalonia and provided the background to all the other historical events that were taking place concurrently. The peasants were tied to the land they worked and had no legal right to abandon it. Despite being allowed to care for a farm, they had to swear allegiance to their lords, pay taxes, relinquish a part of their harvests and serve the lord if called upon.

The Hostoles valley in La Garrotxa has the privilege of being the centre and main instigator of what was probable the first organised peasant revolt in Europe. The confrontation between the numerous peasants and the tough and inflexible feudal structure gave rise to a number of outstanding figures including, amongst the peasants, the almost legendary figure of Francesc de Verntallat who, originally from a well-off family in La Vall de Bas, went from peasant to general under the reign of Joan II.

The term 'Mountain' was commonly used in documents of the era, both in royal missives and in those of the Catalan government (the Generalitat). The 'Mountain' consisted of the Bas, Hostoles and Bianya valleys, Santa Pau, as well as the lands along the rivers Fluvià and Llémena, and the highest uplands of La Garrotxa, all of which had suffered greatly from the impositions of the feudal era. The Mountain was a bastion and permanent refuge of the peasant revolt, and it influenced decisively the development and outcome of the conflict. In 1485, in the aftermath of the defeat of the most radical of the peasant leaders, Pere Joan Sala, King Ferran started negotiations with Verntallat in an attempt to broker a settlement between peasants and lords. An example of this is the fact that on 8 June 1485 representatives of the king arrived in the Mountain and met with Verntallat in his home in Sant Feliu de Pallerols, and demanded that he summon the headmen from the Mountain to a council in the castle of Sant Gregori on 10 June. Francesc de Verntallat was before, during and after the revolt the key to its success. There is no doubt as to his stature as a historical figure given how he was able to act as commander of large-scale social conflict. By 1462 he was the unquestioned leader of the peasants, recognised as such by the traditionally suspicious peasants, the queen, who awarded him custody of the royal flags, and his enemies, who conceded that he was the undisputed head of the uprising.

The sixteenth century was a century of growth. Although serfdom had disappeared from rural Catalonia, the hierarchical social regime remained entrenched. Society was still divided and Catalonia was dominated by an oligarchy of nobles, clergy and large bourgeois *rentiers*.

OLOT grew in size significantly with the end of the plagues and a large influx of French immigrants. The economy also grew considerably, based, above all, on artisan production in general, and wool production in particular. This flourishing of the city was reflected in its urban spread: Carrer Sant Rafael, Carrer Major and Plaça Major were all built up and the encircling city walls had to be expanded. Likewise, the churches of Sant Esteve and Our Lady of El Tura were enlarged.

The predominant artistic style of the age was fruit of the Renaissance, and in Olot the most outstanding constructions dating from this time were the HOSPITAL OF SANT JAUME (mid-sixteenth century), of which a Renaissance doorway survives, and the cloister of EL CARME (sixteenth century) (Fig. 4.7). This cloister was built by Llàtzer Cisterna with money from a legacy left by the Olot notary Miquel Març. Demographic growth continued until the end of the century but was halted by a further outbreak of plague in Olot. The city's councillors hence decided to adopt measures to guarantee the spirituality of the city's inhabitants and improve their education, and so reached an agreement to bring a Carmelite brotherhood to Olot to build a church and school.

The seventeenth century, with Philip IV as King of Spain, was hit by a new cycle of famine, epidemics (more outbreaks of plague), social unrest and wars. In the Catalan countryside one of the main recurrent themes was banditry, perpetrated by poor peasants or even nobles who had been marginalised from the court and political life. These 'bandits' included simple rural thieves, highwaymen and even members of organised

Fig. 4.7 Cloister of El Carme. *Credit* Elisenda Guitart. *Source* Tosca

armed groups (the Nyerros and Cadells) who fought their own private wars. Provoked by the billeting of Spanish troops on the peasantry, in 1640 a revolt broke out; part of the Thirty Years War, this uprising led the Catalan institutions to resist and oppose the attempts of the Spanish monarchy to unify their kingdom via the imposition of the so-called 'Union of Arms' promoted by Count-Duke of Olivares.

During the war between Spain and France, Catalonia backed the French king, Louis XIII, and welcomed him as Count of Barcelona. However, abuses committed by the French troops in Catalonia encouraged the Spanish king, Philip IV, to promise immunity to anyone who changed sides. This century also saw the signing of the Treaty of the Pyrenees, with which the Spanish Crown lost Rossellon and the other counties of northern Catalonia, the Conflent, Vallespir and Cerdanya, which came under the control of the French monarchy. In La Garrotxa, the city of Olot was occupied by French troops for five months.

The eighteenth century was a period of contrasts. Politically, defeat in 1714 in the War of Succession led to a loss of the Catalan institutions of self-government and the 'Castilisation' of Catalan official life (i.e. the Nova Planta decree). This process marked the birth of a unified Spanish state ruled from Castile and the imposition of the absolutism of the new Bourbon dynasty and king, Philip V. Even so, this was also a century of great economic and social gains based on spectacular demographic and agricultural (e.g. specialisation and the opening of new markets) growth, as well as the first incipient steps towards industrialisation and increasingly thriving trade.

Agriculture grew based on the introduction of new products such maize and potatoes and the increase in the available arable land surface due to, for example, the opening up and clearing of the Bosc de Tosca.

In OLOT, the manufacture of stockings and hats (including the *barretina*, the typical headgear of the Catalan peasantry) were the main products of the region's textile industry, which employed mainly women and children. Also highly relevant at this time was the production of 'indiennes', the printed cotton or linen fabrics that were to become the mainstay of the first phase of the industrialisation of Catalonia. These fabrics, manufactured in small workshops or in homes, were produced by stamping prepared cloths with blocks engraved with geometric, historical or plant motifs. Existing buildings originally used by other artisans were adopted for use as textile workshops, and were equipped with tables for the stamping or printing, and boilers for dyeing and colouring.

The manufacture of the indiennes consisted of four phases: spinning, weaving, bleaching and printing. The first, second and fourth phases took place in small factories and workshops in the towns, whilst the third had to be carried out in the open air since it required large amounts of water and space. Before being stamped, the cloths had to be wetted and left to dry in the sun for weeks on end to bleach them; in Olot, this took place in La Moixina in the so-called 'PRAT D'INDIANES'—the 'indienne fields'.

The city's prosperity was reflected in the numerous architectural projects that were carried out during this century including the construction of locally important buildings such as the HOSPICI (1779–1784), the reconstruction of the churches of Our Lady of El Tura and Sant Esteve, and plans to remodel and partially rebuild part of the city. The city council promoted the building of public fountains in a Baroque style (e.g. FONT DE L'ÀNGEL and FONT DEL CONILL) (Fig. 4.8) that had two parts, one designed for the human population and one as a drinking trough for cattle.

The Baroque was a European artistic movement that originated at the end of the sixteenth century and endured into the eighteenth century. It was a response to the crisis of the two essential dogmas of the Renaissance: the admiration for and trust in the ancients, and the optimism of rationalism. 'Baroque' became to be used in a pejorative sense by the neo-classicists of the eighteenth century when referring to the art of the previous century, which they considered to be anti-classical (out-of-proportion, over-burdened, confused and lacking in either balance or order). The most outstanding Baroque monuments were the churches, palaces and private houses; the role of the palace was to reflect the triumph of the absolute monarchies, whilst the Baroque houses were the first manifestation of the noble and bourgeois desire to flee the city and live in the country, not as a peasant or a member of the landed gentry, but as a refined city-dweller.

Baroque churches were characterised by floor plans that, despite maintaining the basilical shape of its predecessors, were more centralised and increasingly complex in design

Fig. 4.8 Font del angel square. *Credit* Llorenç Planagumà

(circular, elliptical, octagonal or even in the shape of a Greek cross). Decorative and symbolic motifs and constructive functions were all blended together into the design of the columns and pillars, which reached its most extravagant form in the spiral solomonic columns that epitomised the solidity and the omnipotence of the Catholic church. A good example of this in La Garrotxa is the ROSER ALTARPIECE in the church of SANT ESTEVE in Olot. The most distinctive element of the roof is the dome, which covers and illuminates the building and symbolises a stretching out to the heavens. Baroque façades were also highly significant as they sought integration into their urban surroundings. Thus, these frontages often bore no relation to the interior of the church and instead formed part of the external urban design project whose aim was to reach out to the faithful. The main doorway was placed at the centre of the façade and was generally extravagant—typically, a porch adorned with a pediment or other ornamental elements, and decorated with columns, niches and corbels. Façades were often concave or convex and in profile created a blend of light and shade that reinforced their visual impact.

Baroque sculpture was characterised by movement, extremes of facial expressions, theatrical gestures and crude depictions of reality. The conception of the human form differed greatly from classical Graeco-Roman portrayals; the principal objective was realism, hence the expressivity of Baroque figures, which was accomplished by depictions of bodily contortions and forced postures, and the representation of movement. Good examples are the life-like sculptures in the NATIVITY SCENES created by RAMON ANDREU during his stay in Olot in 1809–1814.

Baroque painting is characterised by the realism of its rendering of people; great importance was placed on the light that highlighted figures against dark backgrounds. The diagonal lines that create instability and add movement to a canvas were accentuated.

4.5 Contemporary Period

At the end of the nineteenth century, the political and social order of the western world was shaken by a series of events known as the crisis of the Ancient Regime that continued into the nineteenth century. The French Revolution and Napoleon's exploits greatly affected Catalonia. During the so-called 'Great War' (1793–1795), Catalonia was one of the battlefields of the confrontation between the new Spanish monarchy and the nascent French Republic. In 1808, Catalonia and the rest of the Iberian Peninsula were invaded by Napoleon's troops, which marked the onset of the Peninsula War (1808–1814). These wars accelerated the disintegration of the Spanish Monarchy and *ancien régime* in Spain, leading to a paralysation of Catalan economic growth and a delay in its incipient industrial revolution.

OLOT was badly affected and the French occupation in 1812–1814 had serious demographic, economic and social repercussions; similar effects were felt as a result of the dynastic struggles played out in the three Carline Wars of 1833–1840, 1840–1849 and 1872–1876 that brought strife into the heart of Spain and, above all, Catalonia.

The Carlist movement was born on 31 December 1833 when Ferdinand VII died leaving no male heirs to the throne, thereby unleashing a dynastic dispute between the supporters of Isabella II, his daughter and theoretical successor to the throne, and his brother Archduke Carlos. Ferran had abolished the Salic Law to enable his daughter to ascend to the throne in detriment to Carlos; however, unease amongst conservative sections of Spanish society, reinforced by the tensions and unrest generated by the Liberal Triennium and the absolute monarchy, led to the creation of a movement supporting Carlos' efforts to be crowned king.

The recently restored WATCHTOWERS on the VOLCANO MONTSACOPA and the TORRE DE CANADELL (Serra de Vivers) date from this period. La Garrotxa was also heavily influenced by the laws that prompted the dissolution of the monasteries: this legislation, approved under the Liberal governments, decreed that the properties of dissolved ecclesiastical communities be confiscated and put up for public auction.

The BOSC DE TOSCA, an area of fields and woodland near Olot that stands on the lava flow that issued from the volcano Croscat, had been the property of the monastery of Sant Benet de Bages but, with the dissolution of the monasteries, its lands were bestowed on the people who had painstakingly cleared them for cultivation. The clearing of the 2.5 km^2 of the Bosc de Tosca was complete by the end of the nineteenth century and this forested area became a labyrinth of small pastures and fields, dry-stonewalls, tracks and paths.

In the period 1860–1920, the steam-driven factories, concentrated in the main cities, became the main driving force behind the first industrialisation of Catalonia. However, the industrial colonies and factories, situated along the country's main rivers, also played a key role in the consolidation of Catalan industry in the period up to the beginning of the twentieth century. This change in scenery, from the main cities to the interior, was provoked by, amongst other factors, the business class' growing fear of the growth of urban working-class organisation, and the expansion of the railway network into the hinterland of Catalonia. Catalan society was changing rapidly: a dynamic bourgeoisie was assuming a dominant economic role, just as a new, increasingly numerous industrial proletariat was emerging in the cities.

In the final decades of the century, Olot enjoyed great cultural vitality epitomised by the OLOT SCHOOL OF ART. Founded by Josep Berga i Boix and Joaquim Vayreda i Vila, this landscape school of art was both a school of painters and a way of looking at the world based on God, tradition and a sense of nationhood. Its views on both art and life were deeply anti-liberal and represented a reaction to the recurrent social and political agitation of the second half of the nineteenth century. The members of the Olot School were painters, writers and politicians. Artistically, its importance lies in its members' depictions of the landscapes around Olot, their chosen subject matter. This landscape was portrayed as a placid, harmonious paradise, awash with verdant, luxuriant nature replete with meadows, natural springs and marshlands. Their art also depicted traditional festivities, domestic scenes and customs, all set in an orderly, well-manicured landscape where healthy peasants laboured; Olot's distance from the metropolis helped foment this idealisation of the region and convert it into a symbol of peace and prosperity.

Soon, this art would find a faithful public in Barcelona and this pictorial—and also literary—recreation of Olot helped encourage numerous people from the Catalan capital—some in industry, others in art or just art-lovers—to start to spend their summers in the area. Olot and its surroundings became idealised and eventually a romantic, semi-mythical idea of the city grew up—but only at the same time as it began to flounder due to the excess of repetitive and clichéd works painted by the School's followers, and to the death in 1895 of Joaquim Vayreda.

THE SAINTS WORKSHOPS were the most important industry in La Garrotxa in the nineteenth century. 'Arte Cristiano' (which is still operating), the first workshop manufacturing religious images to be opened in Olot, was established at the end of the nineteenth century by Josep Berga i Boix, Valentí Carrera and Joaquim Vayreda, three local artists who decided to set up a society to give work to the students who had studied at the Olot School of Art. In the twentieth century other saints workshops such as El Sagrado Corazón, Las Artes Religiosas and El Arte Olotense sprung up, often opened by the workers themselves. During the twentieth century this industry underwent periods of expansion (the first two decades), stagnation (during the Republic and Civil War) and crisis (in the aftermath of the Second Vatican Council). The Latin-American market has often been able to compensate for the progressive decline in the demand for religious art in Spain and in Catalonia. However, today the manufacture of saints and other religious images is only a small-scale activity with a future that is, at best, uncertain.

The first decades of the twentieth century were years of great political instability and severe social strife. The call-up of reservists to go to fight in Morocco was the last straw for many of the Barcelona working class who were already up in arms and struggling to obtain better social and working conditions. A general strike was declared in protest, which derived into a veritable popular revolt—the Setmana Tràgica (1909)—that was unmercifully repressed. Also of significance was the strike in La Canadenca (1919), the electric company that supplied 60 % of Catalonia's electrical power, that forced a temporary shutdown of 70 % of Catalan industries.

Great social unrest occurred in SANT JOAN LES FONTS, OLOT and the other towns in the county of La Garrotxa, and there were both strikes and lockouts (the right of a business to close temporarily as a way of responding to the demands of the workers; during a lockout the workers do not get paid).

The so-called 'strike of the 21 weeks' was called in OLOT to press for a reduction in daily working hours and improvements in salaries and working conditions. The textile workers laid down tools first, and were followed by workers in the building trade and the manufacture of religious images (many of whom were laid off). The resulting lockouts condemned many Olot families to poverty and it was necessary to create the 'Ranxo al Torín', a place where the people could go and get a square meal.

4.5.1 Modernisme

At the end of the nineteenth and beginning of the twentieth centuries, the artistic movement of *Modernisme* began to take root in Olot (Fig. 4.9). It developed in parallel to similar currents in most European countries—for example, Art Nouveau in France and Belgium, Sezession in Austria, Jugendstil in Germany, and the Glasgow and Chicago Schools—and was known as Modernismo in Spain and throughout the Castilian-speaking world. Although these styles had a series of characteristics in common that provide a clear continuum, each was blessed with its own particular traits and peculiarities.

Modernisme embraces most forms of artistic expression, from painting, graphic design, sculpture, literature to jewelry design. However, it was its architecture and its complementary disciplines, from sculpture to furniture making and all aspects of interior design, that formed a body of work of extraordinary significance. Here, we look specifically at this type of architecture and the skills and crafts that were associated with it.

Fig. 4.9 Gaietà Vila House, Modernist building. *Credit* Octavi Bonet. *Source* Tosca

Barcelona was the centre of Modernist architecture and from here it spread throughout Catalonia and to Mallorca, Valencia and other parts of Spain, often thanks to commissions received by Catalan architects or architects who had studied in Barcelona. Modernism became very popular in Catalonia and, alongside the great architects of renown and their masterworks, was employed by many architects in their designs for buildings throughout the country. At times it was mixed with other styles or applied only superficially—the so-called 'Modernisme de Façana' (façada)—as a series of decorative adornments; this is the case of the *trencadís* (the creation of mosaics from pottery shards), which became exceptionally popular in the whole of Catalonia. Modernism took deep root in Catalan society and endured even once the nation's avant-garde architects had embraced Noucentisme, the artistic current that superseded it.

Catalan Modernism and the considerations of architecture of the age must be viewed as part of the European movement that Catalan architects had studied in books and art and architecture magazines, and seen on their travels.

The desire to develop a new architecture always exists—a new style that reflects contemporary events and uses novel materials and building techniques, and takes into account new social realities in order to move on from the old eclectic and historicist ideas of architecture. Such change does not aim to supplant the past but, instead, to use it as a starting point; in this sense the article published by the architect Domènech i Montaner in the magazine La Renaixença entitled *In search of a national architecture* is significant as it mirrors other writings published in other countries expressing similar approaches.

Finally, it is important to note that Modernisme cannot be understood without reference to Romanticism. It developed on a basis of apparently contradictory ideas: the desire to create a new style of architecture that would fit in with into European currents but would also become a Catalan national architecture; and the willingness to use new materials and techniques (e.g. iron and glass) whilst recouping old artisanal ways of working.

Schematically, Catalan Modernism is characterised by a series of recognisable features: the importance of colour, achieved using stucco, sgraffito and ceramic dressings, and exposed materials; the expressivity of the materials used based on their natural colour and textures, and the contrasts between materials such as stone, bricks, wood, ceramics and iron; the expressivity of constructive solutions and elements, including undressed brickwork and stone, with carefully executed and exposed constructive details including visible metal structures; great decorative richness applied to any element in the building, including sgraffito, moulded ceramic reliefs, tile and *trencadís* adornments, the use of metal such as wrought or moulded iron and tin, stained glass, stone or stucco sculptures, floral and plant motifs, and curved lines—the *coup de fouet*—and shapes.

Forms of this nature were used right from the beginning of the artistic movement, which reveals the debt it owes to the Arts & Crafts movement that used them in its graphic and furniture design. There was also a return to the use of geometric shapes, influenced by the Sezession and Glasgow School, as well as a desire to create complete 'total' works—the architect would not merely create the 'walls' but would also take care of the most minimum details of the interior of the building such as the balcony railings, the door-handles, the floor surfaces, wainscoting, lighting and the furniture. This meant that the architect had to work closely with a whole series of artisans including sculptors, painters, potters, locksmiths and glass blowers. Asymmetry, movement and complex volumes provided the Modernist architects with the emotions they were looking to provoke. They based their work on medieval and, above all, neo-Gothic criteria, perhaps because they believed that theirs was an art capable of elevating the spirit. This was the moment in which nations such as Catalonia were forming, each, as the Modernists asserted, with their own particular signs of identity. They upheld the tradition of the artisanal production of unique original pieces, and rejected the serial factory production of objects. Modernism as a style of architecture was individual-based as its works were in many cases commissioned by wealthy individuals and were often exuberantly chromatic and decorative.

Olot possesses a number of different Modernist buildings contracted by rich residents that all possess a same general style: CASA PUJADOR by Josep Azemar (1911), CASA ESCUBÓS by Josep Paluzie (1905), the LATERAL FAÇADE OF THE CHURCH OF SANT ESTEVE by Martí Sureda (1905), CASA GAIETÀ VILA by Alfred Paluzie (1905), CASA SOLÀ-MORALES by Lluís Domènech i Muntaner (1913) and CASA GASSIOT by Alfred Paluzie (1911).

Geosites and Geoitineraries

5

Joan Martí, Llorenç Plangumà and Xavier de Bolós

This chapter of the book offers a general but detailed view of the main features of the volcanism in La Garrotxa Volcanic Field. A number of selected outcrops and viewpoints from the northern (La Garrotxa Volcanic Zone Natural Park) and southern sectors of this volcanic field are described, and indications are given on how to reach them (Fig. 5.1). The aim is to help readers interpret the landscape and the volcanological processes that characterise this volcanic field, and to provide the necessary practical directions for understanding from a geological perspective some of its most representative volcanic features. Additional information on landscape, natural habitats and cultural heritage are also provided in cases where such information is particularly relevant.

While Chap. 3 explains the significance of the presence of volcanoes in this region and the geodynamic framework in which they developed, as well as the origin and composition of the volcanic rocks and their eruptive history, in this chapter we concentrate on describing the examples that best illustrate these features and, in particular, how volcanism has modelled the landscape of the region.

5.1 La Garrotxa Volcanic Zone Natural Park

5.1.1 Castellfollit de La Roca Cliff

From an observation point located on the old road from Olot to Girona, there is an excellent view of the Castellfollit de la Roca basalt cliff, which consists of two superimposed lava flows with well-developed columnar jointing. The village of Castellfollit de la Roca sits atop a volcanic outcrop between the rivers Turonell (to the south) and the Fluvià (to the north) about seven kilometres from the city of Olot. To reach Castellfollit from Girona, take the N-260 road through Banyoles and on past Besalú. The basalt scarp and village can be viewed at 45 km at the junction of the road to Oix. Park here and follow Natural Park itinerary 13 down to the river Fluvià. After 500 m, cross the river on a wooden footbridge and head towards the foot of the cliff for closer views of the basalt columns.

This cliff is the result of the emplacement of two lava flows along riverbeds and their subsequent erosion by the rivers Fluvià and Turonell. This outcrop—50 m in height and 1 km long—reveals the internal structure of a lava flow (Fig. 5.2). It has been eroding away for thousands of years, mostly as a consequence of erosion by the river Fluvià; however, the process of frost weathering (freezing-thawing), which is all the more effective given the existing jointing, is also relevant. The cracks in the cliff are weak points where weathering is concentrated, which eventually leads to the crumbling and fall of the blocks. They are then carried off by the river's periodic spates, thereby preventing the fallen blocks from building up and stabilising at the bottom of the cliff.

The base of the cliff consists of layers of Eocene sandstone and marl lying underneath gravel composed of many limestone, sandstone and, exceptionally, basalt pebbles. On top, there is a 40-m-thick layer of black and grey basalt. About nine metres from the top of the volcanic materials, a layer (0.2–1.5-m thick) of clay and pyroclasts, easily recognisable by the abundant herbaceous plants that grow there, divides the escarpment into two discrete parts.

a. The lower part has three clearly differentiated layers: (1) the first (starting from the bottom) is 5.5-m thick and has columnar jointing with prisms around 50 cm in diameter; however, it is often hidden by the riparian vegetation; (2) the second has lenticular jointing and is 3.5-m thick; (3) the final

J. Martí (✉) · X. de Bolós
Institute of Earth Sciences Jaume Almera, CSIC, Barcelona, Spain
e-mail: joan.marti@ictja.csic.es

X. de Bolós
e-mail: xavier.bolos@gmail.com

L. Plangumà
Tosca, Enironmental Services, Olot, Spain
e-mail: llplanaguma@gmail.com

J. Martí and L. Planagumà (eds.), *La Garrotxa Volcanic Field of Northeast Spain*, Geoheritage, Geoparks and Geotourism, DOI 10.1007/978-3-319-42080-6_5

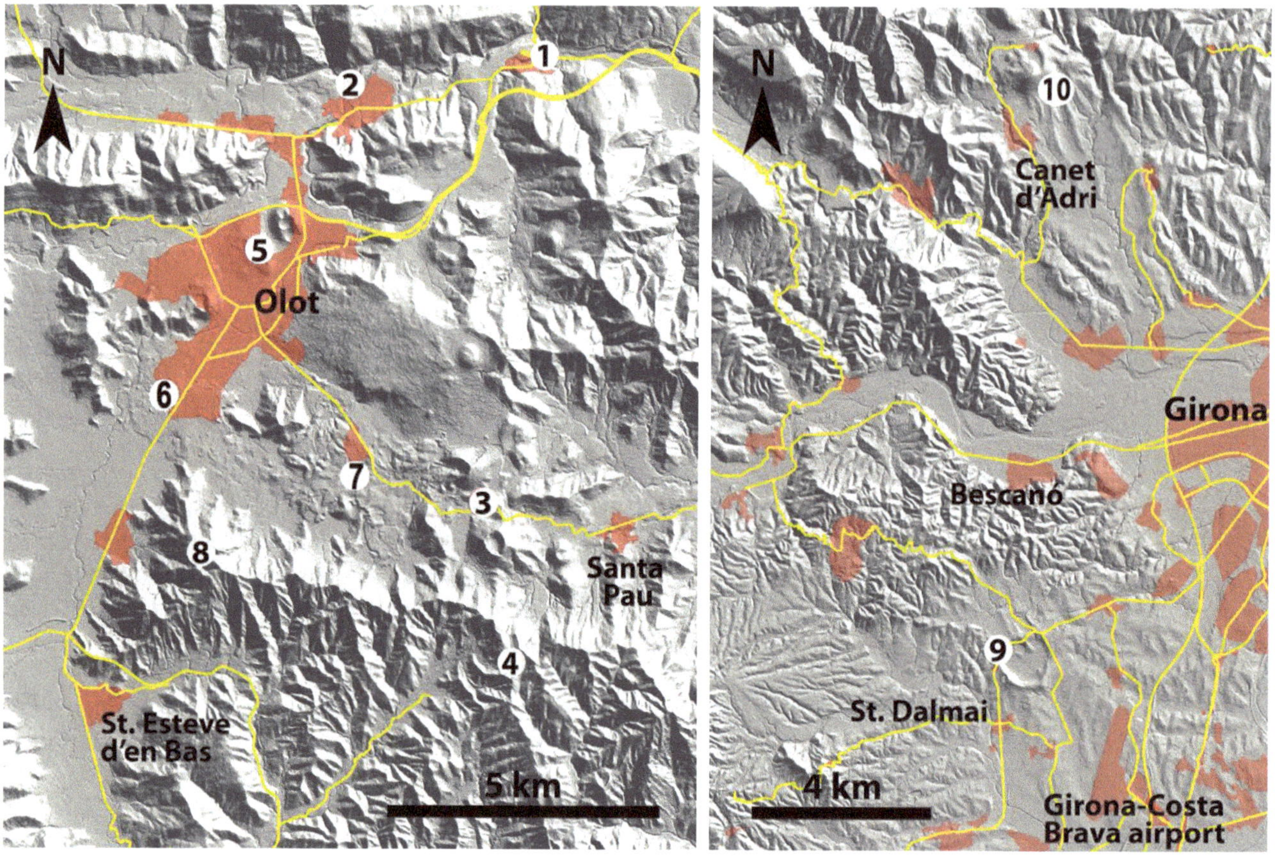

Fig. 5.1 Location of the geosites described in this chapter. *1* Castellfollit de la Roca cliff, *2* El Boscarró quarry, *3* Croscat and Santa Margarida Volcanoes, *4* Can Tià volcanic outcrop, *5* volcano of Montsacopa, *6* lava flow in the Fageda d'en Jordà, *7* Bosc de Tosca with the rootless volcanoes, *8* View from Xenacs, *9* Volcano of Sant Dalmai, *10* Volcano of Puig d'Adri

layer is less than a metre thick and again has columnar jointing but with columns that are only 30 cm in diameter.

b. The upper part has four layers: (1–3) the first three are 5–9-m thick and display obvious columnar jointing; (4) the final layer near the top is about 9-m thick and has well-developed spheroidal weathering.

Around 217,000 years ago a lava flow from the volcanoes on the Batet plateau flowed down and then along the old Fluvià valley to beyond where now stands the town of Sant Jaume de Llierca. Then, some 192,000 years ago, a second lava flow emitted by the volcanoes around Begudà flowed down the Turonell valley as far as Castellfollit de la Roca. In both cases differential cooling of the lava gave rise to a succession of layers in its interior. The time elapsed between these two lava flows was marked by the development of a soil and the build-up of sedimentary materials, which form the deposit that is clearly visible between these two flows. The rivers Fluvià and Turonell overcame these obstructions to their courses once they had begun to erode away the contact zone between the basalt materials and the sedimentary rocks.

The medieval village of Castellfollit de la Roca was constructed on top of these lava flows using, above all, rocks and stones from the flows themselves as building materials.

5.1.2 Sant Joan Les Fonts Columnar-Jointed Lava Flows

This geosite consists of three separate outcrops near the town of Sant Joan les Fonts that exhibit different aspects of basaltic lava flows: El Boscarró (Fig. 5.3), a former basalt quarry that was abandoned in early twenty-first century, stands above the right bank of the Riera de Bianya; the Fontfreda cliff quarry lies alongside the same *riera* (stream); finally, on the left bank of the river Fluvià at Molí Fondo, water erosion has uncovered another sequence of lava flows.

To reach Sant Joan les Fonts from Olot, take the GI-522 road towards La Canya or, if coming from Girona along the N-260, turn right towards Sant Joan just after Castellfollit de la Roca. To get to the outcrops on foot, follow Natural Park

Fig. 5.2 Photograph of the village of Castellfollit de la Roca. *Credit* Pep Callis

itinerary 16 from the main square in Sant Joan, which is also where you should park.

The Riera de Bianya flows into the river Fluvià at Sant Joan les Fonts. Erosion by these watercourses has uncovered three superimposed lava flows that run along the former beds of these rivers. The extraction of basalt rocks from quarries in the early twentieth century enables us today to interpret the relationships between these three lava flows and their internal structures, and to reconstruct their emplacement history.

At El Boscarró different types of jointing are visible in the most recent of the three lava flows emplaced along the Fluvià valley. Five layers can be distinguished here: the lowest has columnar jointing with 5- or 6-sided columns, 20–40 cm in diameter and 2–3-m high. The second and fourth layers have slab jointing, while between them in the third layer there is massive material with a few cooling cracks. The fifth and final layer, just below the soil level, is far more altered due to its proximity to the surface and has a marked spheroidal structure. North-west from the quarry face, the Riera de Bianya follows the contact zone between the volcanic materials and the reddish Eocene sedimentary materials.

At El Molí Fondo, the first and oldest lava flow lies on the bed of the river Fluvià. To the right of the weir, which was built on this first lava flow, the blue-grey basaltic lava exhibits a degree of columnar jointing. If you walk downstream along the left bank of the river, the slabs underfoot correspond to the base of the second lava flow. In places rocks stick out and reveal the rough cinder base. On the cliff next to riverbed the remains of this second lava flow with columnar jointing is visible. On top of this lava flow there is a layer of sediment consisting of sandstone and basalt pebbles in a silt matrix. Finally, the third Boscarró lava flow sits on top of this alluvial layer.

The Fontfreda cliffs correspond to the third lava flow, the same one as is visible at El Boscarró. The lowest layer has clear columnar jointing with columns that are over three metres in height, and an area of lenticular jointing above. Unlike at El Boscarró, the transition between these two layers is here obvious.

The geological history of these three outcrops can be deduced from the emplacement of the lava flows. The first lava flow, emitted by the volcanoes on the Batet plateau, followed the old bed of the river Fluvià and filled in part of its basin. However, the river's erosive power gouged out a

Fig. 5.3 Photograph of El Boscarró quarry. *Credit* Pep Callís. *Source* Documentation Centre, Garrotxa Volcanic Zone Natural Park

new riverbed in the lava flow at the same time as it deposited sediments on its surface. Many thousands of years later, the riverbed was occupied by a second lava flow whose origin has not yet been determined. Over time, sedimentary materials (silt, sand and pebbles) were deposited on top of this second lava flow and once again grew to form a river terrace. Finally, about 133,000 years ago a third lava flow from the volcano of La Garrinada was emplaced on top of these fluvial sediments. This lava flow ground to a halt just past the town of Sant Joan les Fonts.

El Molí Fondo was once a paper mill and its history is closely linked to that of the town of Sant Joan. Although the factory dates back to 1723, its most important transformation took place in 1841, when it was purchased by Pere Capdevila. He undertook a large series of reforms and introduced continuous paper milling that signalled the start of the factory's industrial production. Inside El Molí Fondo there is a permanent exhibition that tells the story of this important example of the industrial heritage of the county of La Garrotxa.

5.1.3 The Volcanoes of Croscat and Santa Margarida

The volcanoes of Croscat and Santa Margarida (Fig. 5.4) are the most significant edifices in the northern sector of La Garrotxa Volcanic Field and hence also in the La Garrotxa Volcanic Zone Natural Park. Croscat lies halfway between Olot and Santa Pau in a relatively flat area, with the Corb-Finestres mountain ridge to the south, Sant Julià del Mont to the north-east and the Batet basalt plateau to the north. The quarry on its northern flank is an exceptional site that reveals the internal structure of this cinder cone. To reach the information centre (Can Passavent) next to the quarry, take the GI-524 road from Olot towards Santa Pau. After seven kilometres, park in the Àrea de Santa Margarida car park on the right. From here, follow Natural Park itinerary 15 to Can Passavent.

In operation from the late 1950s to the early 1990s, the quarrying carried out in the flank of Croscat has revealed an about 150-m-high and 500-m-wide outcrop of pyroclastic materials (Fig. 5.5). On the right-hand side, the quarrying was executed in great terraced steps to help stabilise the volcanic material. However, the middle and the opposite side are less stable and landslips are more frequent. The different layers of scoria made up of irregular, highly vesicular juvenile fragments, for the most part lapilli-sized, are easy to spot. The slope of these layers increases in gradient from the centre to the outside of the cone. The alternation of layers is concentrated at the base of the sequence, where bombs are more abundant. The materials are mostly dark grey or black, although in the area closest to the centre of the edifice they are reddish and ochre. In the lowest part of the quarry there is a reddish layer of welded scoria, which corresponds to the first episode in the cone-building phase of the volcano. If you follow the path back to the information centre, on top of the succession of black Strombolian pyroclastic deposits there is a 2-m-thick laminated layer of finer pale-brownish material that corresponds to phreatomagmatic material deposited at the end of the volcano's explosive activity.

Back at the Santa Margarida car park, take Natural Park itinerary 4 and after 200 m turn right up towards the crater of Santa Margarida. However, to best understand this volcano's structure and materials, continue straight on along the road to Mas El Cros and into the eastern sector of the volcano. Pyroclastic deposits outcrop on the right-hand side of the road, the best sequence of which appears 400 m past the junction.

Santa Margarida is a phreatomagmatic volcano that sits on Eocene rocks. Its crater is circular, about 350-m wide and

Fig. 5.4 Photograph of the Croscat and Santa Margarida volcanoes. *Credit* Joan Martí

70-m deep. Its cone is not formed entirely of volcanic materials, however, and its southern inner rim contains pre-volcanic rocks due to the fact that its crater is embedded into the substrate. In the middle of the crater stands a Romanesque chapel, which has been heavily restored in more recent periods.

On the outcrop along the road to Mas El Cros three types of volcanic materials, sloping successively from right to left, are visible. On top of a silty soil, which corresponds to the pre-volcanic substrate, lies a layer of compacted ash. On top of this there is a layer of black juvenile fragments and fairly rounded reddish-brown lithics. Next, there is a layer of lithic and juvenile lapilli-sized fragments, which predominate. They are black and slightly rounded in shape and have little vesiculation; the lithics are mostly red sandstone. Finally, at the top of the sequence there is a deposit—a fine-grained scoria deposit with no stratification—that closely resembles the previous layer but without any lithics.

The volcanoes of Croscat and Santa Margarida represent two different phases of the same eruption. Along with the much smaller cone of La Pomareda, located a couple hundred metres north-east of the base of Croscat, they lie along a 3-km-long eruption fissure oriented NW-SW. The eruption started at the southern end of the fissure with a vent-opening phreatomagmatic phase that created the relatively large (350-m wide and 70-m deep) explosion crater of Santa Margarida on top of the Eocene basement. This first eruptive phase generated a massive lithic-rich pyroclastic flow deposit, visible on the eastern flank of Santa Margarida, along with several widespread beds of medium-to-coarse-grained, dilute pyroclastic surges and associated fine-ash deposits, which covered most of the area and formed the unit on which the Croscat sequence was built up. This phreatomagmatic phase was followed by a short Strombolian phase that generated a thin, lithic-rich, coarse lapilli fallout deposit that overlaid the previous deposits in the vicinity of the crater. Immediately after these first two phases, the eruption progressed through the central and northern sectors of the fissure extruding basaltic magma and generating massive spatter and occasionally rheomorphic-welded scoria agglomerates, which formed the first cone-building deposits of Croscat and La Pomareda. No more magma was emitted during this and the subsequent phases from the Santa Margarida crater. Thereafter, the eruption was concentrated in the central part of the fissure and changed from fissural (Hawaiian) to central conduit

Fig. 5.5 Photograph of the Croscat quarry. *Credit* Pep Callis

(Strombolian) activity, and the rest of the Croscat cinder cone was constructed.

The Strombolian activity in Croscat generated two main lapilli fallout units: the lower one overlies conformably the basal spatter cones and consists of a several-metres thick, poorly stratified, coarse lapilli deposit with several bomb and scoria beds; the upper unit constitutes most of the volume of the cone and is composed of a thick (several tens of metres), well-stratified-to-thinly-laminated, medium-to-fine-grained lapilli deposit with a few bombs and scoria fragments. This upper lapilli unit also corresponds to most of the intermediate-to-distal outcrops lying to the east of the volcano that are identifiable at distances over 5 km. This unit also covers the Pomareda scoria and spatter cones, as well as the phreatomagmatic deposits and basement of Santa Margarida, which explains this latter volcano's false aspect of a cinder cone. At the top of the upper lapilli unit there is a phreatomagmatic deposit that extends for several kilometres to the east, which changes from planar to cross-bedded stratification in the proximal to distal facies. In its final eruptive phase, Croscat emitted a lava flow, not visible from the main quarry, that breached its western flank. It covered an area of 5 km^2, has an average thickness of more than 10 m and travelled west for 10 km. The total volume of magma (DRE) emitted during the Croscat and Santa Margarida eruption was in the order of 0.2 km^3.

5.1.4 Can Tià

The farm of Can Tià lends its name to the volcano close to the crest of the Corb-Lleixeres mountain ridge whose vent lies at the head of the Sant Iscle de Colltort valley. To get there, first park opposite the restaurant of Can Xel at the 5-km point on the GI-524 Olot-Santa Pau road. From here it is roughly an hour's walk along Natural Park itinerary 5, which takes you straight to Can Tià. On the climb, you pass over Eocene sediments, the upper part of which consists of red and brown sandstones corresponding to the Bellmunt and Folgueroles formations, respectively. Once at Can Tià, the circular shape of the crater and part of the deposits that form the cone, exposed in the quarry next to the farm, are immediately obvious.

The volcano of Can Tià is a maar-type construction, with a 270-m-wide and 20-m-deep crater and a flat bottom. It is asymmetrical—its lip is higher to the south—and the sequence of materials that built this volcano, observable in the quarry next to Can Tia, consist only of pyroclastic deposits (Fig. 5.6). However, there are four different units from base to top. The lowest unit corresponds to a 10-cm-thick, internally massive pyroclastic surge deposit composed of disperse and small (<1 cm) juvenile and lithic clasts, immersed in a highly altered ashy matrix, which rest unconformably on a palaeosoil. The second unit is formed by a poorly stratified, non-welded, highly vesiculated scoria and coarse lapilli deposit, up to 6-m thick, with a few lithic clasts, some of block size. The third unit overlies conformably the previous one and consists of a 1.5-m-thick, thinly laminated, well-sorted, fine-grained lapilli deposit rich in lithic clasts of red Eocene sandstones of variable size (<2–30 cm), with an interbedded ash layer in its upper part. The number of lithics increases gradually towards the top of this deposit, which is sheared by the emplacement of the uppermost unit. This final unit is a massive pyroclastic-flow deposit, up to 3-m thick, that contains abundant Eocene lithic clasts and scoria, and highly vesiculated lapilli fragments, all enveloped in a lithic-rich, ash matrix that has been almost completely transformed into clay minerals, zeolites and iron oxides.

The sequence of deposits displayed in this volcano reveals that its explosive activity began with a short phreatomagmatic event, which immediately changed into a magmatic event (second and third units) and then back into a phreatomagmatic event (upper part of the third unit and fourth unit) over the course of the eruption. The sequence of deposits and the distribution of the lithics show that the first change in eruption behaviour (from phreatomagmatic to magmatic) was abrupt but that the second (from magmatic to phreatomagmatic) was much more gradual. As well, the nature of the lithics found in these deposits clearly indicates a variation in the position of the fragmentation level during the eruption. Although all the lithics correspond to Eocene rocks from the basement below the volcano, they are concentrated differently in each phase of the eruption. The first unit contains mostly lithics from the Bellmunt Formation, which constitutes the main aquifer in the area and is located

Fig. 5.6 Photograph of the Can Tià volcanic outcrop. *Credit* Pep Callís. *Source* Documentation Centre, Garrotxa Volcanic Zone Natural Park

about 300 m below the surface. The lithics found at the base of the second unit are grey sandstones from the Rocacorba Formation, the uppermost stratigraphic unit in this sector of the study area. Towards the upper part of this unit, there is a significant proportion of lithics of brown sandstones from the Folgueroles Formation. The third unit is characterised by the progressive appearance of red sandstone lithics belonging to the Bellmunt Formation, located deeper in the stratigraphic sequence of the area. These lithics become clearly predominant towards the top of this unit and constitute the main lithic fraction in the fourth unit, which represents the culmination of the second phreatomagmatic phase and marks the end of the eruption.

5.1.5 Montsacopa

Montsacopa is one of the five volcanoes that stand inside the city of Olot. It rises right in the heart of the city, between the volcanoes of La Garrinada to the north-east and Montolivet to the south-west; on its summit stands the chapel of Sant Francesc, built in the nineteenth century, and two watchtowers. La Garrotxa Volcanic Zone Natural Park itinerary 17 starts at the Volcano Museum, crosses the whole city and climbs to the top of this volcano. To shorten the walk, you can park in the Olot cemetery at the bottom of the cone, next to the quarry, and then walk up to the crater.

This volcano consists of a single, regular-shaped cinder cone (Fig. 5.7). The construction of this cone involved an initial Strombolian phase and a final phreatomagmatic phase whose deposits are well exposed in the quarry behind the cemetery. The succession of deposits here is composed of a lower, 20-m-thick dark unit formed by well-stratified scoria with alternating layers and different grain sizes, some including centimetric-sized bombs. It is topped off by a 15-m-thick upper brown-to-pale unit formed of thinly stratified pyroclastic surges and fallout deposits with a high content of lithic clasts. This succession of deposits reveals that the eruption started and progressed for a while in a purely magmatic fashion but changed into a phreatomagmatic eruption when the magma interacted with the shallow aquifer located in the unconsolidated Quaternary sediments. This contact produced several explosion breccias, as well as both diluted and concentrated pyroclastic density current deposits, and represented the final phase of the explosive activity.

Fig. 5.7 Photograph of the Montsacopa volcano. *Credit* Eduard Masdeu

A walk around the crater lip gives excellent views of La Garrinada, Montolivet and Bisaroques, three of the other volcanoes in Olot. To the north-east stand the three craters of La Garrinada: at the base of this volcano the first crater, part of a tuff ring that formed during a phreatomagmatic phase, is visible. This crater is almost completely covered by the cinder cone constructed during the subsequent Strombolian phases, which also gave rise to the other two craters that appear on top of the volcanic edifice, one on the southern side and the other on the northern side. Montolivet lies to the south-west of Olot and consists of a cinder cone embedded in the slopes of the mountain of La Pinya with a crater opening north-eastwards. Bisaroques to the south-east sits on the northern slopes of the Batet plateau and, like Montolivet, has a horseshoe-shaped crater. Judging by the deposits found here, although there must have been some phreatomagmatic phases during its eruption, its cinder cone was formed during a Strombolian phase. This cone was partially destroyed in the final stages of its activity by a small lava flow that emplaced northwards to where today we find the bed of the river Fluvià.

5.1.6 La Fageda d'en Jordà

This beech forest lies on the plain formed mainly by the lava flow that was emitted by the volcano Croscat (Fig. 5.8). Today, it is a natural reserve, part of La Garrotxa Volcanic Zone Natural Park, and covers over 200 ha stretching from the Olot-Santa Pau road to the foot of the Serra del Corb ridge. This area also includes the volcano of Puig Jordà and a group of lagoons, Basses d'en Jordà, that are included in the Inventory of Catalan Wetlands.

La Fageda d'en Jordà is one of the best-known and most-visited sites in the natural park. It stands at 550 m a.s.l. on the lava flow emitted by Croscat and is characterised by its undulating relief, full of hillocks of varying sizes that are known locally as *tossols*.

The interest of this beech forest lies mainly in the fact that it stands on relatively flat terrain, on a lava flow and at quite low altitude. The substrate is basalt, which forms when a lava flow cools. The deciduous forest that grows here contains two main habitats: beech with green hellebore where the soil is deepest (andosol) and beech with the tussock grass *Descampsia* sp. on the more acid terrain of the *tossols*. The colours change as the year progresses and, above all in autumn, are one of the main attractions of this forest.

The *tossols* in La Fageda d'en Jordà were formed when the lava flow came into contact with an area of marshland. The heat of the lava (1000 °C) evaporated the water and the rising steam deformed and eventually broke the crust of the partially solidified lava flow. The results are these hillocks, today covered with basalt fragments. It has always been impossible to cultivate this area and it known technically as a 'malpais'—thus, remarkably, this area's inhospitality has saved its forest from being cut down.

Fig. 5.8 Photograph of the lava flow in the Fageda d'en Jordà. *Credit* Octavi Bonet

Natural Park itinerary 2 is circular and begins outside the information centre at Can Serra (Fageda d'en Jordà). Begin by walking through the tunnel under the Olot-Sant Pau road to where on your right stands the monolith engraved with the famous poem by Joan Maragall celebrating the Fageda d'en Jordà.

From here, a set of steps leads down into the beech forest. Natural Park itinerary 2 here coincides with itinerary 1, waymarked in red. When you reach a large *tossol*, the two routes separate. Itinerary 1 continues straight on (it is worth continuing along this route a little way as far as the old terraced fields of can Falet), while itinerary 2 turns left, completes a full circuit of this *tossol* and brings you back to this junction.

5.1.7 Bosc de Tosca: Rootless Volcanoes and Tumuli Lava Field

La Garrotxa volcanic zone has around 40 superficial lava flows that for the most part are *aa* in morphology (i.e. malpais with rough surfaces and block lavas), although there are some *pahoehoe* morphologies (smooth lava flows) in areas such as Batet de la Serra. Of these 40 lava flows, only five have morphologies that can be associated with the *tossols*, and all have three elements in common: they all erupted in periods of much colder climate than the current period, extend over a relatively flat area due to the barrage effect of other lava flows, and are less than 25,000 years old. Other lava flows with *tossols* may exist but could have been eroded away or buried underneath other lava flows or sedimentary deposits generated by the barrage effect that lava flows have on local rivers. In all, 230 *tossols* have been identified and just on the Bosc de Tosca lava flow—situated between the volcano of Puig Cabrioler and the river Fluvià—there are around 110 *tossols*. As well, *tossols* have formed all along the lava flow emitted by the volcano of Puig Jordà that has been dated at the Fonts de Sant Roc as being around 17,000-years old.

The lava flow on which the Fageda d'en Jordà stands was generated by effusive emissions from the fissural eruption that formed the volcanoes of Croscat, Sant Margarida, Turó de Can Xel and several small associated spatter cones. The eruptive activity had different explosive Strombolian and phreatomagmatic phases, and a final effusive phase caused by the degasification of the magma. This effusive phase gave rise to an *aa*-type lava flow that spread across the marshy plain created by the barrage effect of older lava flows in the area. This lava flow is characterized by a profusion of small hillocks, of which 103 are *tossols*. Most are distributed in the distal part of the lava flow, furthest away from the crater. The origin of this area of marshland can be found in the barrage effect of previous lava flows. The Montolivet lava flow crosses and forms the plain known today as Pla de Dalt. The date of this eruption is unknown, although a relative date has been calculated based on the fossil fauna found under a layer of pyroclasts thought to have originated from this volcano. This fauna was typical of the region 9500–18,000 years ago. This lava flow spread across a flat valley generated by the barrage effect of the different lava flows that exist in and around Olot. In all, seven *tossols* have been identified that had already been mapped at the beginning of the ni nineteenth century. The structure of these *tossols* is very similar to those in the Bosc de Tosca.

Both the origin and age of the Parc Nou-La Moixina lava flow are unknown. It lies partially underneath the Puig Jordà lava flow and thus must be at least 17,000-years old. In all, 10 *tossols* have been catalogued from this lava flow.

Scientifically, *tossols* should be referred to as 'rootless volcanoes'. The name refers to the fact that, rather than corresponding to a fissural conduit through which magma rises and provokes an eruptive episode, they are the product of the interaction between a lava flow and an area of humid sediments. *Tossols* are an uncommon phenomenon on Earth, only appearing on very few lava flows, and to date have been described from Iceland and the northern United States. Curiously, certain morphological formations on frozen sediments on the planet Mars have also been described as *tossols*, a discovery that suggests that the conditions needed to form them are highly complex.

They form when a lava flow runs over a wetland or area of marshes. The humid sediments heat up as they come into contact with the lava and generate steam. The steam then rises up under the lava flow through the fractures that appear as the lava cools and provokes small-scale eruptions that deform the surface of the lava flow.

On all of the studied lava flows, the *tossols* are a material resource that has been quarried for the rocks they contain. This provided local inhabitants with a good supply of stone that could be worked into blocks that were lighter than other local rocks due to the vesiculation (large pores) present in these volcanic rocks.

The first attempts to build using rocks of this sort date from the seventeenth and eighteenth centuries and include the Collell bridge and the enlargements of certain farmhouses permitted by the economic prosperity of the time. These rocks were used almost to the end of the nineteenth century and, for example, the Fonts de Sant Roc (1883) were built from stone extracted from the inside of a *tossol*.

Consequently, most *tossols* were converted into quarries that were then used to supply stone for the building of houses and public buildings. One of the best sites in the area for seeing *tossols* is the Bosc de Tosca (formerly known as Vora Tosca or Malatosquera), an area situated to the south of the city of Olot and north of Les Preses, and part of the La Garrotxa Volcanic Zone Natural Park (Fig. 5.9). Its volcanic nature (it lies on the lava flow emitted by the volcano Puig Jordà) makes it unique in the Catalan countries. The Parc de Pedra occupies a small part of the Bosc de Tosca and consists of a labyrinth of paths, dry-stone walls, small pastures, and stone huts whose aspect changes with the seasons. Via the execution of a European LIFE environmental project, the town council of Les Preses managed to rehabilitate this area, thereby restoring, respecting and consolidating the work that had been carried out there 150 years ago.

5.1.8 Xenacs Landscape Overview

Puig Redon (909 m a.s.l.) stands at the far western end of the Serra del Corb ridge where you will also find the Xenacs

Fig. 5.9 Aerial photograph of Bosc de Tosca with the rootless volcanoes. *Credit* Jaume Vicens. *Source* Documentation Centre, Garrotxa Volcanic Zone Natural Park

Recreation Area, one of the best viewpoints over La Garrotxa volcanic zone. From here there are excellent views over the main axial Pyrenees, the pre-Pyrenees and the sub-Pyrenees, as well as most of the Olot trough and Vall d'en Bas.

To get there, take the C-152 from Olot and, about 300 m after passing through Les Preses, turn left up a minor road. After about 5 km of climb, park in the recreation area's car park. From here, take the signposted path to the Puig Redon viewpoint. The road up is closed to coaches and to other vehicles on weekdays, but you can apply for a permit from Les Preses Town Council. Alternatively, walk up to Xenacs following Natural Park itineraries 10 and 11, which start from Les Preses.

The track to Xenacs affords good views over the Vall d'en Bas (1) (Fig. 5.10). This agricultural plain was once a lake: the lava flow that issued from Croscat ran down to the bed of the river Fluvià and obstructed its course, thereby forming a natural dam and lake. Over time, the sediments caused by the erosion of the surrounding slopes silted up the lake, a process that was culminated in the eighteenth century when the plain and its lagoons and marshes were drained for farming.

From Puig Redon, you can see where Croscat's lava flow spread across the landscape by looking for the wooded areas, mostly corresponding to the D'en Jordà beechwood, that cover this lava flow (2).

On a clear day you can see most of La Garrotxa from Puig Redon, as well as parts of El Ripollès to the west and El Pla de l'Estany and L'Alt Empordà to the east. If you look northwards, you can appreciate:

a. Axial Pyrenees (3): the mountains in the background whose highest peaks, snow-covered for most of the year, are made up of ancient Palaeozoic rocks.

b. Pre-Pyrenees and sub-Pyrenees (L'Alta Garrotxa) (4): chains of mountains between 1000 and 1500 m a.s.l. running south of the Axial Pyrenees. They mostly consist of Eocene

Fig. 5.10 Aerial view from Xenacs with the Pyrenees in the background. *Credit* Pep Callís. *Source* Documentation Centre, Garrotxa Volcanic Zone Natural Park

rocks that were intensely folded and affected by faulting during the Alpine orogeny.

c. The mountains of the Sistema Transversal: these are the closest peaks and include the Serra del Corb (5), and are composed entirely of Eocene rocks. They consist of a series of raised and sunken blocks, the product of a system of normal faults. The Collsacabra mountains to the south-west and Puigsacalm (6) to the west are the highest peaks in this sector.

The low-lying area in the foreground to the north is the Olot trough (7). Also visible are the volcanoes of Montsacopa, La Garrinada, Les Bisaroques, Puig Cabrioler, Puig Astrol, El Pujalòs, Puig de la Garça, Croscat, Puig Jordà, Puig de la Costa and Santa Margarida.

The depression bordered by L'Alta Garrotxa to the north, the Serra del Corb to the south, Sant Julià del Mont to the east and Collsacabra and Puigsacalm to the west is known as the Olot basin or trough. In the background and surrounding this depression of tectonic origin stands the majority of the volcanoes in La Garrotxa.

The valleys are all U-shaped since they were filled in by the lava flows emitted during the volcanic eruptions or by sediments that built up in the lakes formed behind the volcanic barrages. From Puig Redon you can see almost the whole of the northern sector of La Garrotxa volcanic zone, including 14 of the (8) 40 volcanoes in the Natural Park. A characteristic feature of the volcanic edifices is their conical shape with craters that are either circular or horseshoe-shaped. They are covered in woodland and almost

Fig. 5.11 Photograph of the crater of the volcano of Sant Dalmai. *Credit* Pep Callís. *Source* Documentation Centre, Garrotxa Volcanic Zone Natural Park

always stand out from the croplands that encircle their bases. Note too the Batet plateau (9) to the north-east, which is formed from a superimposition of successive lava flows from the region's oldest—and so now most heavily eroded—volcanoes.

5.2 Nearby Geosites

5.2.1 Crosa de Santa Dalmai Maar

The Crosa de Sant Dalmai volcanic edifice corresponds to a typical maar-diatreme volcano, almost completely constructed from phreatomagmatic deposits, that forms a circular tuff-ring surrounded by a shallow crater of about 1.3 km in diameter (Fig. 5.11). It lies between the villages of Aiguaviva, Estanyol and Sant Dalmai, and straddles the borders of the counties of La Selva and El Gironès. From Girona, take the Santa Coloma road (GI-533) through Aiguaviva. About 1 km after the junction with the road to Estanyol, there is an esplanade on the right of the road from where volcanic material used to be quarried. In the part farthest from the road, you can climb a low mound, about 5-m high, formed of pyroclasts, which offers a good view across the Crosa de Sant Dalmai crater. The phreatomagmatic deposits, mostly lithic-rich breccias and fine surge and fallout deposits, can be observed in the quarry located below this observation point.

This volcano is located at the boundary between La Selva tectonic depression, infilled with Pliocene and Quaternary sediments, and the mountains of the Sistema Transversal formed from Palaeozoic granites and metamorphic rocks. The tuff-ring is asymmetrical as it is higher (maximum height: 50 m) in the west where the internal and external slopes are also steeper than in the east (maximum height: 30 m). The deposits surrounding the rim are also asymmetrical in the same sense, and extend further eastwards. The sequence of deposits that form this tuff-ring shows consistent stratigraphy, thereby suggesting that most deposits were radially distributed from the vent up to almost 4 km eastwards and only a few hundred metres westwards. This asymmetry in the distribution of the deposits seems to be related to the differences in the strength of the rocks that form each side of the basement below the volcano. To the east, the basement consists of unconsolidated Pliocene and Quaternary gravels, whereas westwards the country rock is composed of Palaeozoic granites and schists. This difference in rock competence seems to have played a major role during the eruption and to have facilitated the eastward excavating effect of each explosion. The sequence of deposits consists of 30 alternating units of lithic-rich explosion breccias and phreatomagmatic fallout deposits, as well as crudely stratified coarse-grained pyroclastic surge deposits. The eruption ended with a Strombolian episode originating from a new vent that opened in the interior of the maar and gave rise to a small cinder cone and a lava flow emplaced within the maar.

Fig. 5.12 Photograph of the Font de la Torre outcrop. *Credit* Pep Callís. *Source* Documentation Centre, Garrotxa Volcanic Zone Natural Park

5.2.2 Pyroclastic Flow Deposits from Puig d'Adri Volcano

This geosite owes its importance to a particular type of volcanic deposits that are not common in other similar volcanic areas: massive pyroclastic flow deposits that share features with typical siliceous volcanic activity. The best observation point is at Font de la Torre, a natural spring in the village of Canet d'Adri (El Gironès), where two streams, Riera de Rocacorba and Riera de Rissec, converge. From Girona, take the GI-531 road towards the Llémena Valley. About 3 km past the village of Sant Gregori, turn right along the GIV-5313 to Canet d'Adri. Roughly 300 m after Canet d'Adri village centre, turn left towards Mas de la Torre and park next to this house. From here, a track takes you down to the spring on the streambed of Riera de Rocacorba.

Fig. 5.13 Close view of the pyroclastic flow deposit of Font de la Torre outcrop. *Credit* Joan Martí

The volcano of Puig d'Adri lies at the foot of the Rocacorba mountain chain, between the village of Canet d'Adri and the hamlet of Adri. This is the easternmost volcano in the Llémena valley, and stands just seven kilometres from the city of Girona. Three identifiable superposed volcanic edifices were built during different phases of the same eruption. A cinder cone (408 m a.s.l) is the most remarkable edifice and can be easily viewed from behind the church in Canet d'Adri.

Puig d'Adri exhibits one of the most complex eruption sequences, involving five different eruption phases, of any of the volcanoes in this volcanic field. It is located on the Adri normal fault, which brings into contact Palaeocene and Eocene materials and is fossilised to the south by Neogene sediments. The Puig d'Adri eruption began with the building of a 850-m-wide tuff-ring, followed by the development on its western side of a small scoria cone and then by the construction of a new cinder cone—that today forms the main volcanic edifice—covering most of the previous volcanic structures. The eruption started with a phreatomagmatic event that generated an irregularly distributed deposit of explosion breccia and diluted pyroclastic density currents, which emplaced south-eastwards along the main existing gullies and ravines to distances of over 5 km from the vent. The resulting deposits consist of thinly laminated, classical dry pyroclastic surges with high-energy sedimentary structures. This initial explosive phase was then immediately followed by a short Strombolian phase that generated a small scoria and lapilli deposit. The eruption then returned to more intense phreatomagmatic activity and generated a series of pyroclastic surges similar to the previous ones, explosion breccias and a pyroclastic flow that emplaced for over 5 km southwards following the course of Riera de Canet. Most of the tuff-ring was constructed during this second phreatomagmatic episode. The eruption continued with a sustained Strombolian phase, which generated a widespread scoria and lapilli deposit around the main vent that covered most of the proximal phreatomagmatic products and gave rise to the main cinder cone. The eruption ended with an effusive phase that generated two lava flows that breached the north-western flank of the cinder cone, one of which emplaced southwards for more than 12 km. Most of the lithic clasts contained in the phreatomagmatic deposits of Puig d'Adri correspond to red sandstones and marls from the Eocene Bellmunt Formation, which once again indicates the significance of this unit as a regional aquifer. This unit is located several hundred metres below Puig d'Adri.

The pyroclastic flow deposit is best studied at Font de la Torre (Figs. 5.12 and 5.13). Here, it forms a 25-m-thick succession of massive flows units, each 1–5-m thick, with planar contacts, crude columnar jointing and strong induration due to post-emplacement cementation processes. These deposits contain large lithic clasts of Eocene rocks, up to 1 m in diameter, and decimetric juvenile highly vesiculated scoria fragments, immersed in an abundant fine-grained matrix composed of small lithics and juvenile clasts that has been mostly transformed into clay aggregates, zeolites and iron oxides. Lithic clasts tend to show normal grading, while the largest juvenile fragments show reverse grading. Columnar jointing has permitted the vertical erosion of the deposit by pervasive infiltration of meteoric water along the joints.

Cultural and Natural Sites

6

Mireia Tresserras, Xavier Oliver and Llorenç Planagumà

Aside from its geology, La Garrotxa Volcanic Field also boasts many sites of great natural and cultural or historical interest, several of which are described in this chapter (Fig. 6.1).

6.1 Parc Nou

Parc Nou is an urban park that acts as a buffer between the city of Olot and the adjacent countryside. A stroll though the park's botanical garden or a visit to the Volcano Museum provide visitors with a simple way of learning about the natural environment of the Olot basin. The upper storey of the museum building (Casal dels Volcans) houses an information centre of La Garrotxa Volcanic Zone Natural Park, along with the offices of this protected area's management team. This information centre offers personalised attention to visitors looking to find out more about the natural history of the county of La Garrotxa.

This walled park is part of the city's list of catalogued buildings and landscapes. The owner bequeathed the park first to the Hospital of Sant Jaume (1932) and then to Olot City Council (1940), and since 1943 the park has been municipally run. It currently covers 13.11 ha but plans exist to enlarge it south- and south-westwards towards the road to La Deu and the southern ring road.

The scientific value of the park lies as much in its pedunculate oakwood as in its birdlife. Birds of open spaces co-exist with deciduous forest species and all are relatively easy to see. Beetles are another attractive faunal group present here and on summer evenings large stag beetles (*Lucanus cervus*) fly amongst the oak trees. These natural and cultural treasures add to the enjoyment of a walk through the park—perhaps interrupted by a fleeting glimpse of a red squirrel—or a moment of calm stretched out on the grass under the shade of a lime tree.

The Volcano Museum is the natural science section of the county museum of La Garrotxa. It is currently housed in Torre Castanys, an important Art Nouveau building inside Parc Nou, and is the perfect place to begin learning about the natural environment of La Garrotxa and, above all, of La Garrotxa Volcanic Zone Natural Park.

6.1.1 Botanical Garden (Oakwood and Botanical Garden)

The garden is a relict[1] of the humid woodland that once covered much of the Olot basin (Fig. 6.2). It is particularly well known for its wide variety of highly attractive woodland plants that flower in early spring—some of which are very rare in Catalonia—in a number of different habitats (e.g. humid grassland, mixed deciduous woodland and pedunculate oak woodland). The oakwood here consists of magnificent specimens of pedunculate oak (*Quercus robur*), accompanied by box (*Buxus sempervirens*) and holly (*Ilex aquifolium*) bushes, and many have been catalogued since 1990 due to their size (over 20 m in height) and age (250 years). This blend of natural elements confers great scientific interest on the site and more than justifies its preservation.

A herb garden containing many of the plants once used for medicinal purposes by local people and typically grown

M. Tresserras (✉)
Educ'art serveis educatius i culturals s.l, Olot, Spain
e-mail: mireia.tresserras@educart.biz

X. Oliver
Institut Catlà d'Historia Natural, Olot, Spain
e-mail: xevioliver@gmail.com

L. Planagumà
Tosca, Environmental Services, Olot, Spain
e-mail: lloren@tosca.cat

[1]Said of the plants or vegetation that, albeit rare today, were much commoner in past eras.

J. Martí and L. Planagumà (eds.), *La Garrotxa Volcanic Field of Northeast Spain*,
Geoheritage, Geoparks and Geotourism, DOI 10.1007/978-3-319-42080-6_6

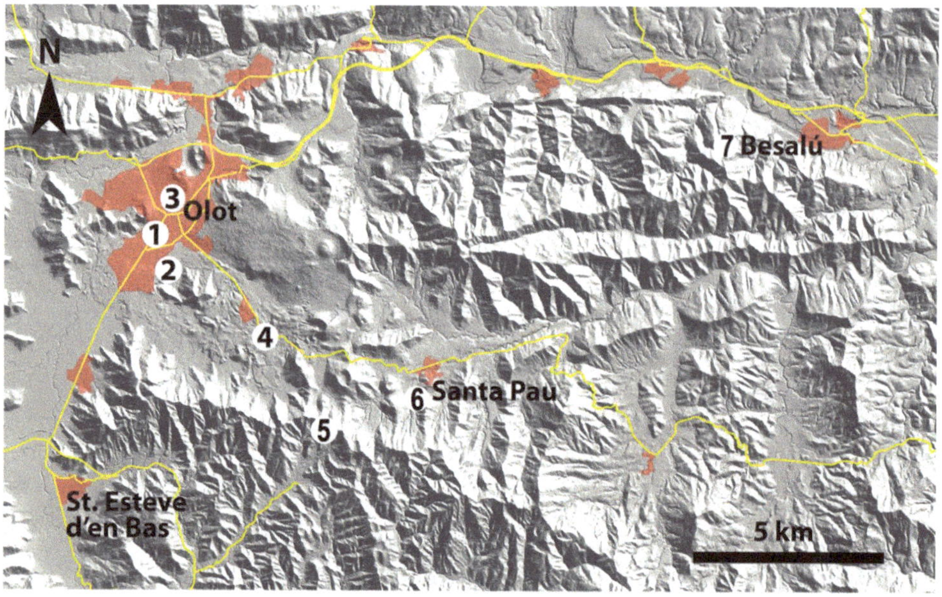

Fig. 6.1 Map of the main points of natural and cultural sites of interest. *1* Parc Nou & Volcano Museum, *2* Springs and woods La Moixina, *3* Olot city, *4* Beech forest Fageda d'en Jordà, *5* Castle od Colltort, *6* The Medieval centre of Santa Pau, *7* Besalú medieval town

Fig. 6.2 Photograph of the pedunculate oakwood in Parc Nou. *Credit* Octavi Bonet

around farmhouses was opened to the public in an area just outside the walls of the park in 2005.

6.1.2 The Casal Dels Volcans, Formerly Known as Torre Castanys

Torre Castanys is today home to the Volcano Museum and the information centre of La Garrotxa Volcanic Zone Natural Park (Fig. 6.3).

The Volcano Museum functions as the natural history section of La Garrotxa county museum and is housed in the lower storey of the Casal dels Volcans. It is an excellent place in which to begin a study of the county's natural environment (climate, geology and ecosystems) and places a special emphasis on subjects of local geological interest such as vulcanism and seismology. One highlight is the audio-visual display on local volcanic and seismic activity that culminates with a simulation of the serious earthquakes that hit Olot and region in the fifteenth century.

6.2 La Moixina

La Moixina, a complex mosaic of ecosystems modified by human activity, is just 15 min on foot from the Casal dels Volcans. It shares with Parc Nou the singularity of its humid oak woodland and the presence in its understorey of a number of flower species that are extremely rare south of the Pyrenees. Another characteristic feature of La Moixina are its wetlands—natural springs and a system of streams that drain into a larger channel, El Rec del Ravell, and the lagoons known as Els Estanys d'en Broc, whose shallow waters harbour water-loving communities of alders, willows, irises, cresses and sedges (Fig. 6.4). However, the geohydrological, botanical and faunistic value of La Moixina is complemented by its importance as a site of historical, scenic and popular interest.

La Moixina is also known as the 'Prat de les Indianes', a name derived from 'indienne', stamped or painted cotton textiles that were printed using wooden blocks engraved with floral and plant motifs. The blocks were inked and then

Fig. 6.3 Photograph of the Casal del Volcans. *Credit* Montse Grabolosa. *Source* Documentation Centre, Garrotxa Volcanic Zone Natural Park

Fig. 6.4 Photograph of the springs and woods at La Moixina. *Credit* Pep Callís. *Source* Documentation Centre, Garrotxa Volcanic Zone Natural Park

applied to the white fabric to create bright, attractive designs. This technique appeared in Olot relatively late on (it had arrived in Barcelona in 1730) when French machines that could work the cotton were first brought to the city. The small artisans who manufactured the indiennes set up their workshops where there was an abundant supply of water, which was essential for the washing and dyeing that was needed to consolidate the colours on the cloth. La Moixina was the chosen site and these workshops operated from 1777 until around 1824, with a maximum of six in operation in 1806. However, when quicker and less painstaking techniques such as the use of synthetic aniline dyes were invented, indienne production began to suffer; nevertheless, two wars, the 'Great' War (1793–1795) and Peninsular War (1808–1814), also played an important part in the decline here and elsewhere of this cottage industry.

Aside from their originality, the importance of the indiennes lies in their relationship with the local art school, L'Escola d'Olot, since to produce the wooden print blocks artisans were required who were skilled in engraving and printing in different colours. The school was opened in 1783 in the Hospice building, with Joan Carles Panyó, an artist from Mataró, as its first director.

When the production of the indiennes came to a halt, Olot was still home to a number of excellent artists, which gave two local painters, Joaquim Vayreda and Josep Berga i Boix, an idea for a new artistic project—the manufacture of religious images. They inaugurated in Olot in 1880 *El Arte Cristiano*, the first factory for manufacturing 'saints', which employed talented local artists that had been retrained in this new type of artwork.

Joaquim Vayreda, Marià Vayreda, Josep Berga i Boix and a whole series of other artists formed the Olot School of Landscape Art, a group of painters who would paint *a plein aire* in places such as La Moixina. Painting in the open air meant taking the easel and other tools out of the studio and painting what the eyes of the artist actually perceived each minute spent painting outside. Consequently, their paintings replicated the sun, its luminosity in that chosen spot, the colours that it accentuated and the reflections that it caused. This would have been impossible in the studio, although there was nothing to stop the artist from adding the final touches to the painting once back indoors if the essence of the moments in the open air had truly been captured. The French painters of the Barbizon School painted in this way and inspired the painters of the Olot School to create their own pictorial style and develop a school of art whose great outdoor studio was La Moixina.

The Vayreda brothers were supporters of the Carlist movement, which meant that they defended tradition and aimed to conserve their positions and status as part of the rural aristocracy. They were staunch defenders of the nation, Catalonia, their land, Olot and its surrounding area, as well as the Catholic religion and its customs. They fought with the Carlines in the Third Carline War and, in the case of Marià Vayreda, left a remarkable testimony in the form of his exceptional novel, *Records de la darrera Carlinada* (Memories of the Last Carline War). For these artists, painting the bucolic landscape of La Moixina, with washerwomen at work and peasants herding the peacefully grazing cows, was a way of bequeathing to future generations their ideal of their country. Although their work was a faithful representation of La Garrotxa, it did not reflect much of the rest of Catalonia at the time. By the 1830s the county had come to be known as the 'Catalan Switzerland' as tourism began to take off here and in other parts of the country.

6.2.1 La Moixina

The site is characterised by a complex landscape mosaic of water-dominated natural systems, profoundly altered by human activity and harbouring a remarkable biological diversity. In addition, its charming landscapes have inspired painters such as the Vayreda brothers and Josep Berga i Boix and so in 1945 was declared a site of National Artistic Importance. Its origins can be found in the lava flow emitted by the volcano of Puig Jordà, which sank underneath the soft alluvial sediments that had formed due to the obstruction by other lava flows of drainage in the Fluvià valley. A series of marshes developed whose base was the aquifer that appears on the surface as a north-south-running line of springs. At the end of the nineteenth century, the area was almost completely drained and the land levelled, and many of the natural streams were converted into artificial channels.

6.2.2 The Pedunculate Oak Woodland (Quercus Robur)

In the past, great pedunculate oak woods and forests would have covered much of the floor of the Olot basin. However, most disappeared with the spread of agriculture and then of the city, and today only occupy 6.8 % of the natural park's surface area. Just three well-preserved patches of pedunculate woodland remain (Parc Nou in Olot, La Moixina and Tossols-Basil) whose floral diversity matches the great oak forests of central Europe. This oak woodland is also one of the areas in the natural park with the richest bird communities.

6.2.3 La Font de La Moixina

Next to the eponymous restaurant stands the natural spring of La Moixina. On private property but open to the public, this natural spring is classified as a Site of Local Cultural Interest due to its historical, scenic and popular importance. Its waters gush from six spouts located beneath an Art Nouveau engraving dating from 1890. It is set in a series of wooded gardens, where stands of tall oaks take root on small *tossols* (rootless volcanoes) interspersed with partially gardened, overgrown spaces in which people from the city of Olot come to stroll in the evening.

6.2.4 La Deu Vella and the Deu Marshes

Also known as La Deu Grossa, this spring is municipally owned and, like its counterpart La Font de la Moixina, is classified as a Site of Local Cultural Interest due to its historical, scenic and popular importance. This permanent and copious spring gushes out from beneath the old wall of the Mas farmhouse, and feeds the attractive nearby marshes of La Deu. Here, visitors can walk amongst the alders, poplars and willows, and alongside the small pools covered by carpets of duckweed (*Lemna* sp.) and watercress (*Nasturtium officinale*), and lined by yellow flag irises (*Iris pseudocorus*), bulrushes (*Typha* sp.) and sedges and rushes (*Juncus* sp.).

6.2.5 La Font de Bufaganyes

This private spring, also classified as a Site of Local Cultural Interest due to its historical, scenic and popular importance, dates from the eighteenth century. Its waters gush from four spouts into a drinking trough and are much appreciated by local people.

6.2.6 Basses d'en Broc

These lagoons were dug in the 1990s in an attempt to recreate the wetland habitats that were once common in the Olot basin. Together with the Deu Vella marshes, they are included in the Catalan Inventory of Wetlands due to their great biological diversity and the presence of several singular species of plant and animal. They attract, above all, numerous species of bird and harbour important hygrophilous plant communities characterised by riparian woodland species such as white poplar (*Populus alba*) and white (*Salix alba*) and crack (*S. fragilis*) willows, together with herbaceous vegetation dominated by reeds (*Phragmites australis*) and bulrushes (*Typha* sp.).

6.3 Olot City Centre

Olot is a city that is always changing (Fig. 6.5). The medieval city was triangular in shape, with the abbot of Ripoll's residence or palace (whose existence survives in the name—Plaça Palau—given to the square that now occupies this site) at its apex. From here, the limits of the city were defined on one side by the wall that ran as far as the church of El Tura—then a Romanesque building situated at the far end of this wall—and on the other by the natural defensive line of the river Fluvià. However, the serious earthquakes that shook Olot in the fifteenth century destroyed much of the medieval city. It would have been too expensive to rebuild on the same site and so, thanks to permission granted by the king Alfons the Magnanimous, local people were allowed to construct outside the city walls.

Thus, the modern city was born with a grid of streets reminiscent of a Renaissance city that runs from the church of El Tura to the church of Sant Esteve. It embraces the city's main square (Plaça Major), onto which all the surrounding perpendicular streets converge. Also in the city centre is the sixteenth-century cloister of the convent of El Carme, designed by the mason Llàtzer Cisterna and one of the few such Renaissance cloisters remaining in Catalonia.

The church of Sant Esteve is today the parish church and old documents talk of a church in Olot that stood on top of a small hill. However, it was not until the peace of the eighteenth century and the onset of a period of growth and

Fig. 6.5 Aerial photograph of Olot. *Source* Tosca, Environment Services

industrialisation that the city could begin to devote time and energy to enlarging its churches and constructing other grandiose buildings such as the Hospici. This three-storeyed building, designed by the Madrid architect Bonaventura Rodríguez, had two cloisters and was built to give shelter to all those with nowhere to eat or sleep. Despite the exponential growth of the city, this building was too large to ever be used as a hospice and has since its inception always been used for other purposes—a fire station, school, Olot's first museum and even as a market place. Also dating from the eighteenth century is Can Trincheria, a splendid town mansion that is currently a museum where visitors can view a huge permanent nativity scene.

By the eighteenth century local industry had grown and Olot boasted important textile and tanning industries, as well as the production of the 'indiennes'. These attractive cotton prints were produced in La Moixina and prompted the establishment of the Olot School of Art (1783), which, today housed in the former convent of El Carme, would go on to train artists for over 200 years. The Olot School of Landscape Art made good use of the marvellous landscapes of the Olot region and gave a name to a style of art that became well known throughout Catalonia. Once the production of the indiennes came to an end, the school went on to train sculptors that would manufacture the saints in the workshop of religious imagery that was opened in 1880. In the nineteenth century over 30 such workshops were producing religious images in Olot; today the factory known as *Art Cristià* has been converted into the Museum of Saints, the only such museum in the world, whose task is to explain to visitors how religious images were manufactured in the past and how they are still produced today.

The nineteenth century was the century of the Carline Wars or the Wars of Succession. Olot was an industrial city and as such had a liberal tradition. However, the Carline forces managed to occupy the city towards the end of the century, but not without first having burnt down a number of important buildings such as the Teatre Principal (built in 1769, although its current façade designed by the architect Martí Suera dates from 1904) and the convent of the Caputxins. This was despite the walls (known as the 'Carline' walls) that surrounded the city from the twin watchtowers of the fortress of Sant Francesc perched atop the volcano Montsacopa, down the volcano's flanks as far as the square Plaça Clarà. From here they continued to the banks of the river Fluvià, thereby securing the eastern side of the city. Each gate into the city was flanked by two towers with the main entrance in between. Wars—just like ignorance—destroy and today no evidence apart from the two watchtowers and the fortress that preside over the city from the summit of Montsacopa remains of Olot's city walls.

The twentieth century provided Olot with the perfect opportunity to recover from the convulsions of the previous century. In 1911 the railway line from Girona to Olot was completed, which meant passenger travel, freight transport and a far better connection with the outside world. There was now no need to depend on the horse-drawn coaches to travel to Girona since the train was much quicker (even despite the song that insisted that the 'Olot train leaves when it wants to, arrives when it can'). This was also the century of Art Nouveau, a superb and exquisite artistic style that left its mark on Olot in the form of façades, sculptures, paintings and even the famous posters advertising cigarettes. The new houses of the bourgeoisie, factory owners and liberal professionals were an ideal outlet for the Art Nouveau architects, then still little known and often forced also to work as municipal architects, who could let their imaginations run free at the behest of those they were commissioned by.

The first such house, halfway along the Firal, is Casa Solà Morales (1913–1916), whose façade and extensions were the work of the well-known Barcelona architect Domènech i Muntaner. The sculptures were by Eusebi Arnau—at that time collaborating with Domènech on the construction of the Palau de la Música in Barcelona—and represent two caryatids or nymphs carrying fruits and food, symbols that reflect the prosperity of the Solà Morales family.

Continuing down the Firal, there are a number of Art Nouveau-style façades such as that of Casa Gaietà Vila (1905), instantly recognisable by its exquisite decorative style (Fig. 6.6). It is known as 'The Dragon', in part due to the bookshop of the same name that has opened there, but mainly due to the myriad of sculptured dragons that decorate its facade. Its architect, Alfred Paluzie, also designed the façade of the house of Doctor Gassiot on the corner of the streets of Sant Rafel and Els Dolors.

On the way to the cloisters of El Carme, visitors will first past alongside the church of Sant Esteve with its green and yellow glazed Manisses ceramic adornments, as well as the balustrade with floral motifs that runs all along this side of the church, both the work of Martí Sureda. Continuing past the main square visitors will reach Plaça del Conill, a small square where the Baroque fountain is crowned by a rabbit (*conill* in Catalan means rabbit). Its design was inspired by the rampant rabbit that forms part of the family shield of the Conill family, which can be seen on the house—once outside the old city walls—that stands opposite the fountain.

Next door stands Can Escubós, a building dating from the seventeenth century that was restored in 1905–1907 by Alfred Paluzie. Its wrought-iron railings and glazed ceramic motifs are typical of the city. Opposite is Can Pujador (1911–1912), also the work of a municipal architect, in this case Josep Azemar from Figueres. It too is Art Nouveau in

Fig. 6.6 Photograph of the Gaietà Vila Art Nouveau house. *Credit* Octavi Bonet

design but is very distinctive and possesses many clear stylistic traits. It is a unique work of art produced by an architect who, like all others, was, nevertheless, always subject to the whims, tastes and financial restrictions placed on him by the person for whom he was working.

All the art produced over the years in the county of La Garrotxa and, above all, Olot, is well represented in the county museum in Olot, today located on the third storey of the Hospici. Here visitors will find clues as to how people lived in the region from the eighteenth through to the twentieth centuries, as well as examples of work by artists such as Miquel Blay, Josep Clarà and Leonci Quera from the Olot School of Landscape Art, the first two the first local exponents of Art Nouveau. Clarà was also a key figure in the Noucentisme movement, while the younger Leonci Quera was ahead of his time and left behind a body of work whose abstract paintings and sculptures are much more contemporary in style. Unmissable is the vast painting *The Charge* by Joaquim Casas and the collection of 46 Art Nouveau posters presented as entries to a competition run in 1900 and 1901 to promote a tobacco brand set up by the Olot entrepreneur Manuel de Malagrida in Argentina. With his money made abroad, he returned to Olot and undertook the construction of the Malagrida New Town, much in the style of the garden-city projects then being planned in other parts of Europe. Today, this calm backwater of Olot is easily visitable and is ideal for wandering and appreciating its particular architectural styles.

The city's three museums (county museum of La Garrotxa, Museum of Saints and Volcano Museum), as well as Casa Trincheria, are known collectively as the 'Museums of Olot'. The Volcano Museum is situated in Parc Nou in a Noucentiste-style building, Torre Castanys. This museum is devoted to the physical environment of La Garrotxa, that is, its seismic and volcanic activity and its main ecosystems. The earthquake simulator gives visitors the chance to 'relive' the earthquakes of the fifteenth century that razed the city to the ground.

Points of interest (mentioned above):

(a) Casa Solà Morales
(b) Casa Gaiatà Vila
(c) Sant Esteve parish church
(d) Museum-house Can Trincheria
(e) Can Escubós
(f) Can Pujador
(g) Cloister of El Carme
(h) Church of Mare de Déu del Tura
(i) Hospici (county museum of La Garrotxa)
(j) Museum of Sants
(k) Malagrida New Town
(l) Parc Nou.

6.4 La Fageda d'en Jordà

This beech forest extends across a plain formed largely of the lava flow emitted by the volcano Croscat (Fig. 6.7). A natural reserve (200 ha) and part of La Garrotxa Volcanic Zone Natural Park, it stretches from the Olot-Santa Pau road to the foot of the Serra del Corb. It also includes the volcano of Puig Jordà and the interesting lagoons of Basses de Jordà (included in the Catalan Inventory of Wetland Areas).

Fig. 6.7 Photograph of the Fageda d'en Jordà in autumn. *Credit* Octavi Bonet

This forest contains some of the most characteristic—and most visited—elements of the La Garrotxa Volcanic Zone Natural Park. It grows on a lava flow at an altitude of 550 m a.s.l., and its surface is undulating and broken by many hillocks or hummocks of varying sizes that are known locally as *tossols.*[2]

The interest of this beech forest lies mainly in the fact that it stands on relatively flat terrain and at quite low altitude. As well, its substrate is basalt, the rock that forms when lava solidifies. The deciduous forest that grows on here contains two main habitats: beech with green hellebore where the soil is deepest (andosol) and beech with the tussock grass *Descampsia* sp. on the more acid terrain of the tossols. The colours change as the year progresses and, above all in autumn, are one of the main attractions of this forest.

[2]*Tossols*: these rounded formations consist of 'blisters in the basalt lava flow from the volcano Croscat produced when the lava crossed an area of marshy land. The heat of the lava (1000 °C) evaporated the water and the steam that was generated penetrated the lava and deformed and then broke its partially solidified crust. The resulting mounds are covered by fragments of basalt.

6.5 Colltort Castle

This ruined castle (Fig. 6.8) stands on the rocky heights of Turó de Colltort (845 m a.s.l.) in the Hostoles valley (municipality of Sant Feliu de Pallerols) at the far eastern end of the Serra del Corb. It is first mentioned in 1017 in a document signed by Ermemiri *ex castro de Collo Tortuensi* and was given by Count Ramon Borrell de Barcelona to the Count of Besalú, Bernat Tallaferro, in return for his vassalage. Its jurisdiction passed into hands of the lords of Hostoles and then to the lords of Santa Pau. By the seventeenth century, it formed part of the royal county of Hostoles.

Due to its position on the frontier, this castle was the scene of many conflicts between the counts of Besalú and Girona. It consists of a square tower and outer walls, with foundations that rest on the very rock of the mountain or on large flat slabs of rock. The mortar used in its construction was lime-based and very resistant.

6.6 Santa Pau

The village of Santa Pau (Fig. 6.9) once belonged to the Count of Besalú, the lord who during medieval times controlled the whole of La Garrotxa and part of the neighbouring counties. The summits of the nearby Serra de Finestres marked the boundary with the adjoining lands of the important Count of Girona and so this ridge was an important defensive point—hence the construction of the castle of Finestres. However, when the feudal lord of Finestres—an armed knight—decided that he wanted to live in greater comfort in the valley below, he chose the small hillock of Puig Angul for his new castle and stronghold. In theory, this was a castle whose objective was to keep the peace, as he already possessed a castle for making war.

This is thus the origin of the village of Santa Pau, whose houses cluster around the square castle (begun in the thirteenth century but heavily remodelled in later periods) and act as its outer defensive wall. The workers who built the castle or who worked there and served the lord lived in these houses, which were built like a 'flock of sheep around its shepherd', according to the great Catalan writer, Josep Pla. The development of the village dates from feudal times, an era whose political and social system was totally pyramidal and hierarchical. The people at the base of the pyramid were the vassals or slaves of the feudal lords. Allegiance was sworn to the lord in a ceremony of *vassallatge* that took place in the lord's castle; from here the hierarchy of power continued upwards, the lords side-by-side with the ecclesiastical authorities, through the counts and finally as far as the king and the Pope. The farmers, millers, blacksmiths, masons and shopkeepers were obliged to pay tithes or give

Fig. 6.8 Photograph of the ruins of the castle of Colltort. *Credit* Tosca, Environmental Services

produce to their feudal lord and religious authorities. Thus, the lord's decision to go and live in the village was also a way of pressurising and controlling his subjects. This was the origin of the castle of Santa Pau, which was handed down from the lords of Finestres to the lords of Santa Pau, who, in time, became the barons of Santa Pau and their lands, a barony. When the family was not living in the castle—for these lords were also warriors that participated in more than one raid designed to extend Catalan influence in the Mediterranean—the lord's *procurador* or representative ensured that the castle was well cared for and that the rents and taxes were paid on time. He lived in the house opposite the castle as can be seen by the ornamental decoration over the window of the balcony.

Between the thirteenth and fourteenth centuries, Santa Pau grew and a porched square was built after a royal decree was issued in 1297 giving the village permission to hold a fair on Mondays. The arches protected the market from the rain, which was more prevalent in those days than it is currently. The surrounding houses have three floors: the ground floor was for the animals, whose heat would warm the people living on the first floor, while the second floor was for drying produce and generally had half-arched windows that opened to the outside.

A document dating from 1300 that encouraged people to go and live in Santa Pau in exchange for certain opportunities and privileges gave the village a welcome boost. However, the earthquakes of the fifteenth century destroyed the parish church, Santa Maria dels Arcs, and so in the fifteenth century it was decided to build a new church, also dedicated to Santa Maria. Its bell tower was built in the shape of a defensive tower and matches the architecture of the castle. In the seventeenth century and, above all, in the eighteenth century, Santa Pau began to grow—the village began to expand beyond its old squares and new streets of houses started to spring up. On many such houses the date of construction is engraved on the lintel of the doorway, alongside a drawing representing the trade of the person who lived there.

6.7 Besalu

Although the origins of the town go back to the Iberian tribes, the Celts and Romans, its most important historical period was undoubtedly the medieval era. A document from the Upper Middle Ages (tenth century) talks of a castle built on a strategically important hillock, where today stand the

Fig. 6.9 Photograph of the medieval centre of Santa Pau. *Credit* Pep Callis. *Source* Documentation Centre, Garrotxa Volcanic Zone Natural Park

remains of the canonical church of Santa Maria. The fortunes of town waxed and waned over the years. However, the apogee of its power came in the eleventh century under the rule of Bernat I Tallaferro (990–1020), when the town became an important ecclesiastical centre. One of the most significant aspects of the legacy from that period is the town's Romanesque architecture.

Documents from the thirteenth century describe the Jewish community that inhabited Besalú in that era, although there are references to Jews in Besalú from as far back as the ninth century. The Jewish community was economically very active and worked in many trades; they were also moneylenders, an occupation that Christians were banned from performing. From 1391 onwards, the Jews began to be persecuted in the Spain and consequently many commenced to abandon their homes. In 1415 Pope Benedict XIII obliged Christians and Jews to live separately, from which point the Jews started to live in *calls* or Jewish quarters that were physically isolated from the Christian quarters.

Important buildings:

- The Mikveh. This was where the Jews carried out their ritual baths and its presence is clear evidence of the importance of the Jewish community in Besalú. King Jaume I authorised the construction of a synagogue and the Mikveh was probably associated with it. Built next to the river Fluvià, these baths enabled the faithful to purify both their souls and bodies.
- The old bridge. The first records of this bridge date from the eleventh century. Its current structure is the result of numerous reforms and reconstructions carried out over the centuries.
- The hospice of Sant Julià and the beautiful façade over the main doorway.
- The church of Sant Pere (once part of the town's Benedictine monastery). Worth remarking are the window flanked by two lions on the main façade and the floor plan with an ambulatory that circles the presbytery.

6.8 Volcanic Cuisine

La Garrotxa is famous for the foodstuffs it harvests and produces. They are all influenced by the climate and the volcanic soils—rich and highly fertile—whose influence is palpable in the local cuisine, a blend of nature and culture that meet on a plate. Its products are easily distinguishable

Fig. 6.10 Photograph of a plate of Santa Pau haricot beans and a typical sausage. *Credit* Llorenç Planagumà

and are produced in sufficient number to reach and satisfy the needs of the market.

A group of local restaurants—known as the 'Cuina Volcànica' group—has developed a passionate style of cuisine based on the county's most traditional gastronomic customs. They believe in the need to represent, defend and promote the region and all its agro-alimentary area of influence, and use in their cuisine products that have been harvested, bred or manufactured in a non-industrial fashion in the area (Fig. 6.10). These products are transformed to create an exceptional culinary experience that satisfies even the most demanding of palates. Their cookery is based on traditional recipes but is also creative and daring, and has helped stimulate curiosity and new ideas regarding local cuisine. In light of the richness and the variety of the county's foodstuffs, the aim of the Cuina Volcànica group is to encourage the use of these products as a fully fledged part of the county's culinary tradition, and is today helping to spread the singularities of its home-grown cookery.

A good example are the Santa Pau haricot beans, now with their own European protected designation of origin (PDO). The PDO *Fesols de Santa Pau* protects the seeds of the traditional varieties of haricot bean (*Phaseolus vulgaris* L.) known as Tavella Brisa, Setsemanera and Gra Petit that, be they dried, cooked or conserved, have been produced in the volcanic soils of the county of La Garrotxa. A 'protected designation of origin' is the term used to describe certain products that originate from a particular place, region or—exceptionally—country, whose quality or characteristics are due fundamentally or exclusively to a particular geographical environment, as well as to both natural and human factors. All production, transformation and processing of these products must be carried out totally within a defined geographical region. The most characteristic features of the volcanic cuisine are the Santa Pau haricot beans, buckwheat, *farro* (a type of semolina made from maize four), black turnips, the Vall d'en Bas potato, lamb, beef, charcuterie and sheep's cheese.

The dried haricot beans are characteristically white, slightly shiny and rounded. All the beans of the varieties Tavella Brisa, Setsetmanera and Gra Petit that are harvested in the PDO and matured from mid-August onwards have high protein levels (25 % or more), which fall if they are grown in non-volcanic soils. If produced in the volcanic area, once cooked these three authorised varieties have a very tender skin, low-to-medium flouriness, a characteristic haricot bean taste and no other conflicting tastes. These factors are due to the type of soil and the fact that the beans are dried at moderate temperatures.

7 Educational Programmes

Octavi Bonet

Just over 200 years ago the Olot naturalist Francesc Xavier de Bolòs made known to the scientific community the existence of the volcanic landscapes of La Garrotxa. La Garrotxa volcanic zone is an area of great geological, wildlife and scenic interest, and, despite being a highly humanised region, its inhabitants have over the years ensured the conservation of this natural heritage. However, urban and industrial growth in the 1960s engendered a series of damaging environmental problems including quarrying, construction projects, polluted rivers and a proliferation of illegal rubbish tips that seriously endangered the region's natural values. The need to protect La Garrotxa and its volcanoes provoked a series of public demonstrations that prompted the Catalan Parliament to approve in 1982 a law—the first ever passed in Catalonia designed to protect a natural area—aimed at guaranteeing the protection of the county's landscapes. Its stated objective was to ensure the conservation of the region's geological, botanical and scenic riches—but not at the expense of the area's economic development.

La Garrotxa Volcanic Zone Natural Park is one of the most singular of all protected areas in Catalonia. It is also the best-preserved volcanic area in the Iberian Peninsula and one of the most significant volcanic regions in mainland Europe. However, it is subject to intense human activity, which has to be carefully controlled to ensure that it is compatible with the preservation of the region's natural qualities.

Currently, the protected volcanic area covers 15,308 ha and is home to almost 40,000 people living in 11 different municipalities. Within the park, there are 28 natural reserves with higher degrees of protection.

O. Bonet (✉)
Tosca, Environmental Services, Olot, Spain
e-mail: octavi@tosca.cat

7.1 What Makes La Garrotxa Volcanic Zone Natural Park (PNZVG) Such a Singular Area for Environmental Education?

The singularity of this volcanic zone is due to a series of natural phenomena including the presence of the volcanoes, a variety of micro-climates and a highly diverse flora and fauna, as well as to the way in which human action has modelled the landscape over the centuries.

The volcanic activity that first occurred in La Garrotxa around 700,000 years ago has always been a potent modifier of the local relief, which in turn has profoundly influenced the location of human settlements in the region. This volcanic area boasts almost 40 volcanic cones and around 20 lava flows and, given that the most recent eruption only occurred 10,000 years ago, it is classified as dormant rather than extinct. Its climate is another determinant factor as *Mediterranean* and *Atlantic* micro-climatic areas are found in juxtaposition: the former is characterized by mild winters and hot dry summers, and the latter by rain in summer and thermal inversions in winter that guarantee low temperatures and frequent frosts. The varied relief, convergence of micro-climates and diversity of the geological substrata are the key to explaining the region's great diversity of plant and animal species, as well as the wide range of natural environments in which they can thrive—in other words, in this region *Mediterranean* environments co-exist alongside *central-European*-type habitats.

The landscape (Figs. 7.1 and 7.2), modelled by both volcanic and human activity, is another feature of the region that must be conserved. For many centuries, human settlement in the region was determined by what land could be cultivated or grazed and by which forests could be exploited. This had the effect of converting the local landscape into a harmonious mosaic of fields, pastures and woods that provided sustenance for local families. However, a great change in the landscape and social relationships occurred with the development of artisanal and industrial activities.

J. Martí and L. Planagumà (eds.), *La Garrotxa Volcanic Field of Northeast Spain*,
Geoheritage, Geoparks and Geotourism, DOI 10.1007/978-3-319-42080-6_7

Fig. 7.1 Example of typical landscape from La Garrotxa Landscape. *Credit* Octavi Bonet

7.2 Environmental Education at the Forefront of Sustainability in La Garrotxa

The pedagogical activities carried out in this protected area are based on the following precepts: the use of the environment as a setting and as an educational resource, and the use of methodologies and attitudes as strategies in fieldwork; the receptors of the educational activities' previous knowledge of the area; the adaptation of the content to the receptor group; and the interest and motivation shown by that group.

This characterisation helps provide a clear vision of the different types of visitors to whom the educational or informational activity is addressed (Fig. 7.3), and has fostered the creation of the environmental educational programmes used in this protected area and its area of influence.

The people who in one way or another will receive information regarding the region's values and singularity can be divided into two groups defined in accordance with the aim of the educational activity they undertake and the use that they will make of it: the **local population** (i.e. people who live in the park and its area of influence, who can be subdivided into students, naturalists, professionals/politicians, members of socio-cultural entities and the general public) and the **non-local population** (corresponding to those who do not live in the park or its area of influence, and who can be subdivided into students, tourists, naturalists and scientists).

These two typologies of visitors to the region can be broken down into other groups in terms of their interests and their capacity to grasp informative content:

- **Students** can be regarded as all those who are still receiving some type of formal education. Students from the ***local population*** are those that receive this education in a school or college in the county of La Garrotxa.
- **Non-formal students** are those children and young people who participate in non-formal educational activities provided by entities that collaborate with the park's programmes. The ***local population*** segment of this group consists of the people that live in La Garrotxa.
- **Tourists** are the most varied group of park users and include families, groups of pensioners and members of recreational, sports and cultural entities who visit La Garrotxa to learn more about its volcanic features. In the

Fig. 7.2 Volcanoes of La Garrotxa. *Credit* Octavi Bonet

case of local residents, this group corresponds to the **general public**, that is, similar types of people with similar reasons for visiting but with more specific interests in one particular aspect of the protected area.

- **Naturalists** consist of amateur wildlife enthusiasts, secondary-school teachers, workers in environmental education and university lecturers who, from a variety of different standpoints, are interested in improving their knowledge of certain specific aspects of the region's physical, natural and human environments, and how they are managed.
- **Scientists** are the least numerous of all the groups and include professional and university geologists (or from related disciplines) who are conducting specific research on vulcanism or who aim to discover more about certain aspects of this field of study (Fig. 7.4). In the case of the local population, this group also includes **professionals/politicians** working in territorial management who have a specific interest in acquiring knowledge of the volcanic zone for use in their day-to-day work.

These types of visitors require appropriate educational and informational programmes, which are often run by the Natural Park in collaboration with local entities such as the Olot City Council's Municipal Institute for Education and Young People, La Garrotxa Consortium for Environmental and Public Health (SIGMA), the Olot Foundation for Further Education, or Turisme Garrotxa, the local tourist board.

In order to develop such programmes it is essential to develop strategic objectives that take into account the type of person to whom they are addressed. Hence, a number of different educational programmes have been developed, each with its specific target public. For the ***Local Population*** in the sphere of (1) **formal education** (e.g. activities forming part of the school curriculum or La Garrotxa school agenda), the aims are to foster knowledge of the natural, socio-cultural and economic circumstances of the Natural Park, to promote the objective and rigorous acquisition, interpretation and communication of information, and enhance the ability to define a scale of values regarding the often conflicting interests of local economic development and

Fig. 7.3 Guide at work. *Credit* Octavi Bonet

the preservation of the territory. (2) **Leisure activities** (summer camps and outward-bound programmes) aim to bring local children into closer contact with their region and to improve their knowledge of their natural and social surroundings via activities designed to increase their awareness of the values of the protected area. Professional environmental educators play an important role in (3) **formative activities** (e.g. park guide training course), which endeavour to improve the educational services offered by local companies working in the Natural Park, to assist in the development of stable pedagogical entities, and to provide continuous in-work training. Via the work of (4) **local entities** (e.g. *Visc entre Volcans* programme), local residents are encouraged to get involved with both the work of the park and that of the entities themselves.

In terms of the ***Non-local population*** and the provision of (1) **formal education,** the park's pedagogical programme aims to improve the quality of both guided and non-guided school visits to the protected area, to avoid excessive impact on the environment and the local population, to improve knowledge of the natural, socio-cultural and economic circumstances of the Natural Park, and to promote the objective and rigorous acquisition, interpretation and communication of information (Figs. 7.5 and 7.6). In the sphere of (2) **tourism** (e.g. information centres and points, training courses for tourism professionals), the aim is to satisfy the demand for information, improve communication with visitors, increase environmental awareness amongst visitors, decongest the busiest areas of the Natural Park, and work in coordination with the region's tourist providers to improve the services they offer. It is hoped that this will make tourism more sustainable, improve the quality of visitor experiences and satisfy visitor expectations, whilst attempting to avoid any negative impacts on either the environment, the local population or on the visitors themselves.

Once the typology of the different receptors of the educational and informational activities has been established, the framework in which each group of visitors are to be attended and the information (i.e. the volcano-related content) to be transmitted to groups from each different sphere

Fig. 7.4 Field course on volcanism. *Credit* Octavi Bonet

Fig. 7.5 **a–c** Summer camp *ENNATURA'T*. *Credit* TOSCA

can be defined. Likewise, it is important to follow the guidelines proposed by the European Charter on Sustainable Tourism (CETS).

The CETS makes an explicit reference to educational activities in its Principle 7: *Formation and Education*, which describes strategies aimed to "… educate local residents and visitors with content related to the region's cultural and natural heritage and socio-environmental impacts."

Work with local schools and schools from outside the region, as well as with adults, aimed at improving awareness of and respect for La Garrotxa and its environment is one of the most important tools at the disposal of the Natural Park for ensuring more sustainable use of the protected area and the region in general.

If we centre our view on the framework of these pedagogical tasks, it is important to appreciate that the idea of

Fig. 7.6 a, b Summer camp *AVENTURA'T Credits* Víctor De Paz and Octavi Bonet

basic competences is more of an economic than educational concept (i.e. the notion of e-skills). A skill is the learned capacity to carry out a task with pre-determined results often with the minimum outlay of time, energy or both. Skills can often be divided into domain-general and domain-specific skills. For example, in the domain of work, some general skills would include time management, teamwork and leadership, self-motivation and others, whereas domain-specific skills would be useful only for a certain job. Often, however, skills depend on a number of different factors.

The project set up by the **OECD** (Organization for Economic Co-operation and Development) known as DeSeCo (Definition and Selection of Competencies) defines a competence as: "*The capacity to respond to complex demands and carry out a variety of tasks in an adequate manner. It entails a combination of practical skills, knowledge, motivation, ethical values, attitudes, emotions and other social and behavioural operations that are activated in order to attain the successful outcome of an enterprise.*"

The **EU** has drawn up a proposal (January 2006) for all member states in which it describes basic competences as: "*... a combination of skills, knowledge and attitudes that is appropriate to each context. They are necessary for personal fulfilment and development, as well as for social inclusion, active citizenship and employment. By the end of their compulsory education and training, students must have developed these competences in order to be equipped to lead an adult life, and will have to continue developing, honing and updating these competences throughout their lives as part of a learning process.*"

These considerations have to be transmitted into the world of education. Spanish education law states that the **aim of education** is ensure that children acquire the necessary tools for understanding the world in which they are growing up, which will guide them in their future behaviour. Education should lay the foundations that permit children to grow into adults that are able to act critically in the diverse, plural and constantly changing societies into which they are born.

Thus, competence should be understood as the application of **knowledge**, **skills** and **attitudes** to the resolution of problems in a variety of contexts. In other words, the teaching of **basic competences** implies the development of KNOWLEDGE (*WISDOM*), CAPACITIES (*HOW TO ACT*), SKILLS (*HOW TO BEHAVE*) and ATTITUDES (*HOW TO RESPOND*), and how to make use of all these PERSONALS RESOURCES (*HOW TO REACT*).

To advance in the attaining of these basic competences it is essential to view educational processes as running along four main pathways:

1. Learning to be and to act independently
2. Learning to think and to communicate
3. Learning to discover and take the initiative
4. Learning to cohabitate in the modern world

These processes must be approached via subject matters that can be separated into concepts (basic knowledge of the elements of the physical world and geological processes), procedures (ability to interpret and communicate

Table 7.1 Subject matters used to approach educational processes

Concepts		Procedures		Attitudes	
A	Geomorphological elements: volcanic edifices, lava flows and other significant features	1	Obtain information	I	Evaluation of the exceptionality of volcanic activity
B	Lithology: volcanic materials	2	Communicate information	II	Respect for the environment
C	Location and toponyms	3	Observation and recognition of geomorphological features	III	Acceptance of personal and collective responsibility for the conservation of the environment
D	Magma genesis	4	Interpretation of geological processes connected with volcanic activity	IV	Ability to come up with alternative solutions to environmental problems
E	Magma ascent	5	Orientation	V	Participation in actions aimed at resolving environmental problems
F	Type of eruptive activity				
G	Type of volcanic deposits				
H	Modification of relief features				
I	Uses and impacts of anthropic action on volcanic features				
J	Territorial management: personal action and administrative powers				

Source PNZVG volcanic conservation strategy

information) and attitudes (behaviour, norms and values), as summarised in the table (Table 7.1) on the following page.

Two main blocks of content have been distinguished in this analysis of the pedagogical tools currently available in the Natural Park: the first refers to the local population and the second to the non-local population (Table 7.2).

For each visitor type the details of the content that should be studied are given. Additionally, for each pedagogical product (programmes, activities, facilities/services and resources such as publications) the content currently studied is shown.

7.3 Implementation of Educational Programmes at La Garrotxa Volcanic Zone Natural Park

The graphs shown in Figs. 7.7, 7.8 and 7.9 provide details of the project and the work carried out by La Garrotxa Volcanic Zone Natural Park that over the past four years has begun to be questioned by the Ministry of Agriculture, Animal Husbandry, Fisheries, Food and the Environment (DAMM) of the government of Catalonia.

We analyse quality of environmental education using three different parameters:

(a) the number of hours of work schools devote to the subject before and after visits to the Natural Park;
(b) balanced ratios (1/20);
(c) loyalty.

An emphasis on quality in environmental education services also fosters the stability and correct formation of the educators that work in the Natural Park. Key to the understanding of this process is whether the number of hours schools devote—above all in the classroom—to the subject is increasing or decreasing. If the hours rise, then there is an interest in the subject matter that the school has studied—or will study—in the park; however, a fall in the hours denotes a lack of interest. Thus, the work that students do that is associated with environmental education and the hours they spend on it are key elements for determining the quality of the learning they receive in the park.

In the case of La Garrotxa Volcanic Zone Natural Park (Figs. 7.7 and 7.8), it is clear from the figures that the number of hours spent on the subject matter grew whilst the Catalan government backed the Natural Park's pedagogical programmes. However, once the park and its environmental education came under the prerogative of the DAMM, this interest faded, which is reflected in the growing precariousness of educators' jobs and in the dismantling of the pedagogical programmes that had been operating since 2000. This trend was accentuated by the economic crisis, which meant that school outings—above all those that required a journey—for environmental studies were cut back.

Table 7.2 Main blocks of content according to typology and origin of receivers

Typology	Local population	Non-local population
Tourists	– Informative programmes – Offer of activities for entities – Talks-visits with entities – Workshops on vulcanism – Exchanges with other volcanic areas – Museum of Volcanoes – Network of itineraries-geological sites – Publications (books)	– Informative programmes – Pedagogical services – Network of itineraries-geological sites – Information boards – Exhibition on volcano Croscat – Museum of Volcanoes – General information leaflet – Miscellaneous publications
Students	– Pedagogical activities for local residents – Guided visits—workshops on vulcanism – Exchanges with other volcanic areas – Museum of Volcanoes – Documentation centre – Network of itineraries-geological sites – Publications (books)	– Pedagogical activities for visiting schools – Pedagogical services – Network of itineraries-geological sites – Museum of Volcanoes – Guide for school groups – General information leaflet – Miscellaneous publications
Naturalists	– Offer of activities for entities – Pedagogical activities for local residents—courses/workshops/seminars promoted by NP or in collaboration with other entities – Exchanges with other volcanic areas – Documentation centre – Museum of Volcanoes – Network of itineraries-geological sites – Volcanological map—sig "vulcà" – Publications (books) – Geological site guide*	– Museum of Volcanoes – Geological site guide – Network of itineraries-geological sites – Courses/workshops/seminars promoted by NP or in collaboration with other entities – Exchanges with other volcanic areas – Volcanological map—sig "vulcà" – Publications (books)
Scientists/professionals/politicians	– Exchanges with professionals – Specialist workshops and courses – Natural Park annual report – Volcanological map—sig "vulcà"	– Compendium of scientific articles (Catalan/English), sites and observation points – Guide to vulcanism – Courses/workshops/seminars promoted by NP or in collaboration with other entities – Volcanological map—sig "vulcà"

More hours in the field means that more time can be spent on procedures and attitudes, leaving work on concepts for the classroom. In this way, it is possible to work on aspects that are difficult to address in a closed space: observation, analysis, synthesis, resolution of problems and hypotheses, the drawing of conclusions, and so forth. In other words, more time can be to devoted to more intense work on the basic competences such as maths, linguistics, knowledge, the interaction with the physical world and the use of information.

The maintenance of a maximum ratio of one educator per 20 students allows a close relationship to develop between receptors and educators (teaching-learning). The receptor can participate far more, which in turn can motivate a group and bring the best out of it. The group's interests and curiosities can be explored and this will bring up questions such as what the group members can do to ensure that human action in the park is compatible with the conservation of the territory—not only in the places of greatest value but also those that are closest to the group's homes, be they rural or urban.

A low ratio allows for more intense interaction between participants and also forces participants to think more in any attempt to reach a conclusion, be it as part of a group or on an individual basis. A low ratio also encourages the development of fruitful small-group dynamics, which can involve the exposition of a problem to encourage a group to come to its own conclusions. This type of activity is diametrically opposed to the situation in which the educator controls all the group's activity via a monologue. It is necessary, however, to accept the inherent risks in these kind of group dynamics—as much for the educator as for the receptor—that are the only way of advancing within the confines of the teaching-learning interface (everybody teaches and everybody learns all the time; we all have something to offer). As the ratio falls, the number of hours of study—under normal conditions—increases (Fig. 7.9).

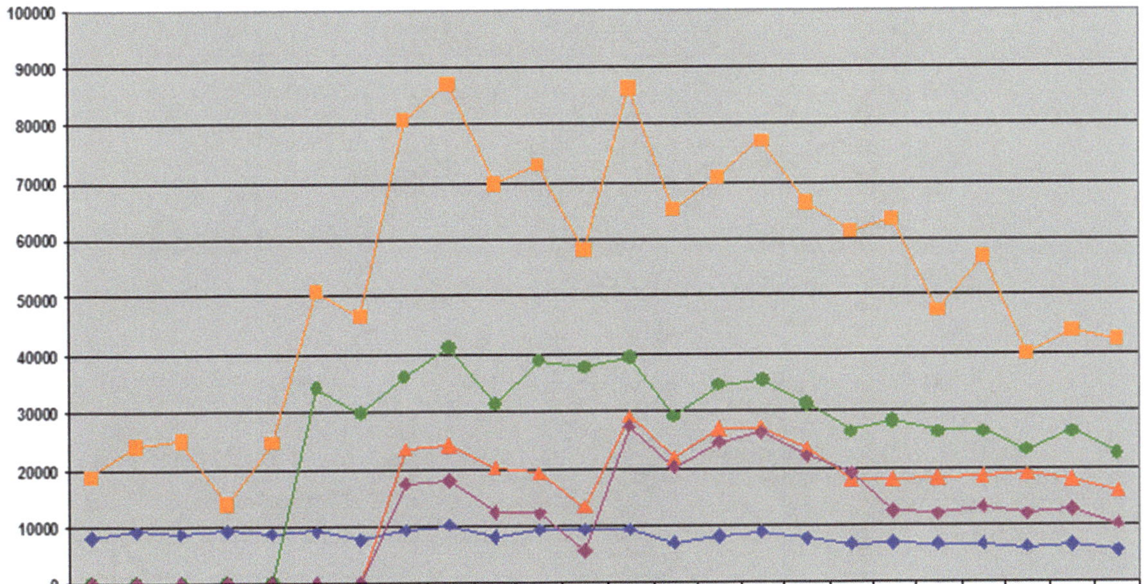

	1991	1992	1993	1994	1995	1996	1997	1998	1999	2000	2001	2002	2003	2004	2005	2006	2007	2008	2009	2010	2011	2012	2013	2014
Total participants	7871	8948	8558	9231	8515	8982	7648	9307	10008	8113	9323	9261	9855	6984	8084	8742	7447	6563	6651	6523	6493	6136	6346	5354
Total hores/participant	18890	24160	24818	13847	24694	51197	4665	80882	87160	6976	73155	58087	86471	65135	70835	77273	66245	61212	63428	47468	56939	39857	44047	42347
Total previ	0	0	0	0	0	0	0	23547	24119	2020	19205	13336	29176	21604	26888	26988	23312	17850	17962	18352	18429	18891	17799	15840
Total camp	0	0	0	0	0	34037	2965	36039	41085	3134	38772	37746	39118	28972	34353	35269	31173	26280	28047	26297	26394	23152	26280	22479
Total posterior	0	0	0	0	0	0	0	17497	17814	1225	12120	5649	27383	20187	24705	26135	21992	18991	12328	11768	13054	11697	12627	10156

Fig. 7.7 Comparative study of hours devoted to environmental education programmes in La Garrotxa Volcanic Zone Natural Park. Total participants; Total hours/participants; Total hours before visit; Total hours in field; Total hours after visit. *Credit* Xavier Oliver and Octavi Bonet

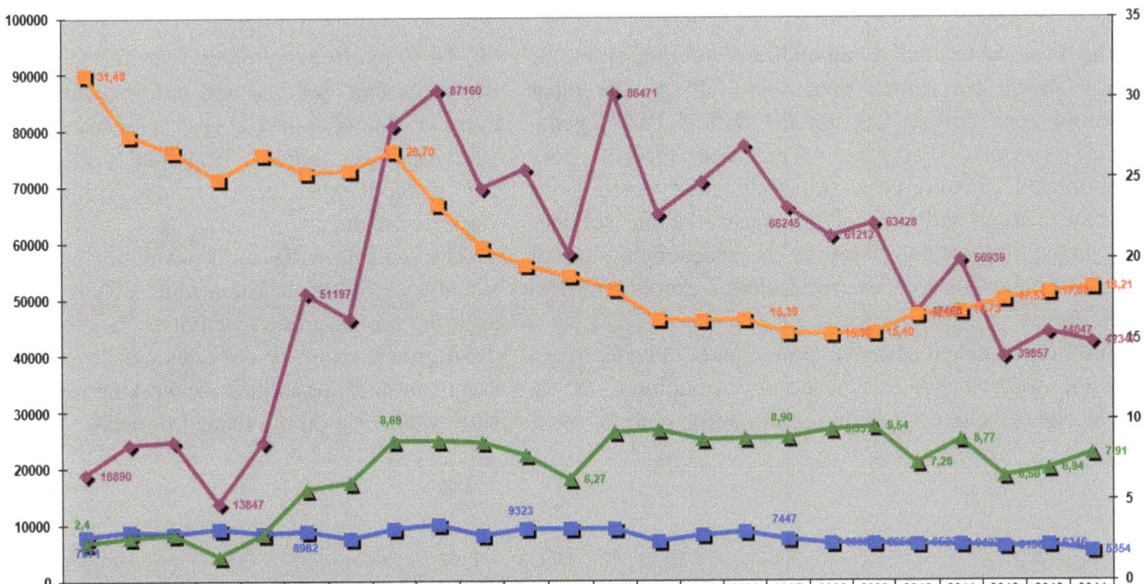

	1991	1992	1993	1994	1995	1996	1997	1998	1999	2000	2001	2002	2003	2004	2005	2006	2007	2008	2009	2010	2011	2012	2013	2014
Total participants	7871	8948	8558	9231	8515	8982	7648	9307	10008	8113	9323	9261	9355	6984	8084	8742	7447	6563	6651	6523	6493	6136	6346	5354
Total hores/participant	18890	24160	24818	13847	24694	51197	46653	80882	87160	69767	73155	58087	86471	65135	70835	77273	66245	61212	63428	47468	56939	39857	44047	42347
Mitjana hores/alumne	2,4	2,7	2,9	1,5	2,9	5,7	6,1	8,69	8,71	8,60	7,85	6,27	9,24	9,33	8,76	8,84	8,90	9,33	9,54	7,28	8,77	6,50	6,94	7,91
Ràtio alumne/educador	31,48	27,70	26,74	25,02	26,53	25,37	25,55	26,70	23,42	20,72	19,60	18,95	18,11	16,22	16,20	16,24	15,39	15,30	15,40	16,51	16,73	17,53	17,88	18,21

Fig. 7.8 Comparative study of hours devoted to formal environmental education programmes in La Garrotxa Volcanic Zone Natural Park: Total participants; Total hours/participants; Average hours/student; Ratio student/educator. *Credit* Xavier Oliver and Octavi Bonet

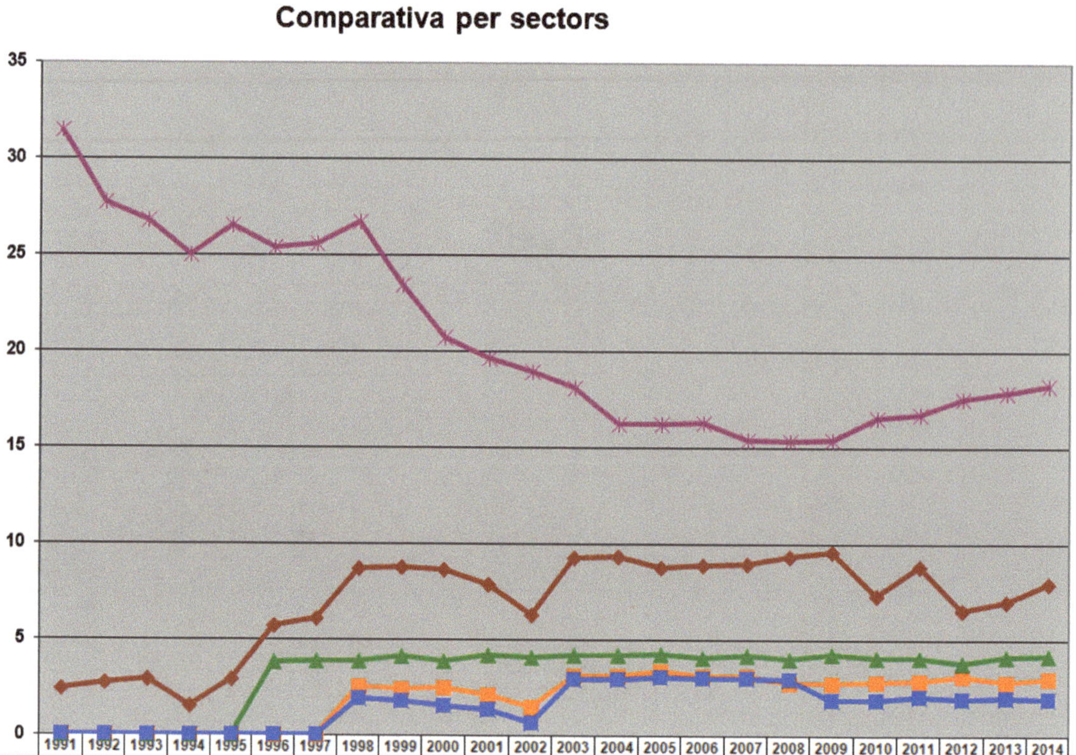

	1991	1992	1993	1994	1995	1996	1997	1998	1999	2000	2001	2002
Mitjana hores/alumne	2,4	2,7	2,9	1,5	2,9	5,7	6,1	8,69	8,71	8,60	7,85	6,27
Mitjana hores/alumne Previ	0	0	0	0	0	0	0	2,53	2,41	2,49	2,06	1,44
Mitjana hores/alumne Camp	0	0	0	0	0	3,79	3,88	3,87	4,11	3,86	4,16	4,08
Mitjana hores/alumne Post	0	0	0	0	0	0	0	1,88	1,78	1,51	1,30	0,61
Ràtio alumne/educador	31,48	27,70	26,74	25,02	26,53	25,37	25,55	26,70	23,42	20,72	19,60	18,95

	2003	2004	2005	2006	2007	2008	2009	2010	2011	2012	2013	2014
Mitjana hores/alumne	9,24	9,33	8,76	8,84	8,90	9,33	9,54	7,28	8,77	6,50	6,94	7,91
Mitjana hores/alumne Previ	3,12	3,09	3,33	3,09	3,13	2,72	2,70	2,81	2,84	3,08	2,80	2,96
Mitjana hores/alumne Camp	4,18	4,15	4,25	4,03	4,19	4,00	4,22	4,03	4,06	3,77	4,14	4,20
Mitjana hores/alumne Post	2,93	2,89	3,06	2,99	2,95	2,89	1,85	1,80	2,01	1,91	1,99	1,90
Ràtio alumne/educador	18,11	16,22	16,20	16,24	15,39	15,30	15,40	16,51	16,73	17,53	17,88	18,21

Fig. 7.9 Comparison by sectors: Average hours/student; Average hours/student before visit; Average hours/student during visit; Average hours/student after visit; Ratio student/educator. *Credit* Xavier Oliver and Octavi Bonet

The third factor that is an indicator of quality is the loyalty shown by schools and, above all, by the other institutions and entities that use the Natural Park's pedagogical services. This factor—provided that it goes hand-in-hand with the other two highlighted factors—can serve as a good indicator of the quality of the services provided. Loyalty alone, however, is not enough, since it may be simply due to economic factors (schools return to where the service was cheapest).

The environmental education programmes in La Garrotxa Volcanic Zone Natural Park have a loyalty rating of 80 %, that is, eight of every 10 entities return every year. Of these 80, 10 % return every other year due to their own internal dynamics (e.g. schools and entities that work on a topic every two years or once every cycle, and so return every two years—primary schools—or three years—pre-schools) or the methodologies they follow (projects, pedagogical workbooks, etc.).

The remaining 20 % do not return or do so only occasionally according to the whims of whoever is charge of deciding on the group's activities. In other words, whether these groups return or not does not depend on the projects that the schools or entities are working on but on the person who is working on the related projects.

8 Geotourism in La Garrotxa

Llorenç Planagumà, Isabel Junquera and Esther Canal

Geotourism is tourism based on geology and promotes geological conservation at the same time as it encourages the economic and social development of local communities. 'Volcanic tourism' is one particular type of geotourism. Volcanoes and volcanic landscapes have a worldwide fascination and many are visited annually by huge numbers of people. Visits to both live and extinct volcanic areas provide public recreation, adventure and enjoyment but also afford opportunities for observing, learning and appreciating the power and role that volcanoes have played in the modelling of the planet's surface (Fig. 8.1). Most protected volcanic areas offer stimulating tourism activities and opportunities for public use, all of which provide substantial community benefits.

In La Garrotxa sustainable tourism is being promoted via the European Charter for Sustainable Tourism, whose principles and strategies aim to promote local sustainable development.

8.1 Economic Activity in the Area

The following data give an idea of the economic activity in La Garrotxa. In terms of the primary sector, animal husbandry has become an intensive industrial activity and the increasingly mechanised local agricultural system is now subordinate to its needs. This has led to the abandoning of marginal lands and the subsequent encroachment of forest cover. Forests now cover 70 % of the total surface area, cultivated land 12.7 % and permanent pastures 8.6 %. The mainstays of secondary industrial production in La Garrotxa are textile production, food processing and metallurgy, which together employ over 60 % of the county's industrial workforce (Fig. 8.2). Finally, tourism is beginning to play a vital role in the tertiary sector of the county's economy.

The municipalities of Les Preses, Olot, Sant Joan les Fonts, Castellfollit de la Roca, Besalú, Sant Feliu de Pallerols and Les Planes d'Hostoles are clearly industrial in nature, while Mieres, St. Aniol de Finestres, Maià de Montcal, Sales de Llierca and Sant Ferriol are all agricultural. Santa Pau, La Vall d'en Bas and Riudaura form a third group in which there is a better balance between the primary and secondary sectors, as well as important recent growth in tourism-related service industry.

8.2 Sustainable Tourism and the Conservation of Areas of Natural Interest

Year after year, the number of visitors to the protected areas of Catalonia rises. Driven by a variety of motives, people are increasingly interested in discovering those areas of the country that have been declared areas of special protection. Around 5.1 million people visited Catalan protected areas in 2012, a figure that rose to 5.5 million in 2013.[1]

Public bodies and the motors of economic development concentrate their efforts on fomenting tourism as an economic activity with an important multiplier effect that can generate both direct and indirect economic benefits, above all in regions that have traditionally been excluded from the usually positive—and very visible—economic dynamics of

L. Planagumà (✉)
Tosca, Environmental Services, Olot, Spain
e-mail: lloren@tosca.cat

I. Junquera · E. Canal
Repte Territorial, Olot, Spain
e-mail: Isabel@repte.net

E. Canal
e-mail: esther@repte.net

[1]Estimated figures according to the number of visitors entering information and visitor centres in Catalan natural parks in 2012 and 2013 published in the reports *Environmental data in Catalonia*. 2012 and 2013. See bibliography. From 2014 onwards, these documents no longer provide any data on visitor numbers in Catalan protected areas.

J. Martí and L. Planagumà (eds.), *La Garrotxa Volcanic Field of Northeast Spain*,
Geoheritage, Geoparks and Geotourism, DOI 10.1007/978-3-319-42080-6_8

Fig. 8.1 Aerial photo of the volcanic zone. *Credit* Octavi Bonet

the tourist sector. These areas, which tend to be visited by fewer tourists despite the protected areas they harbour, are now commonplace objects of tourism promotion campaigns aimed at motivating visitors to come and 'discover' and 'experience' their delights.

In light of this situation, protected areas have increasingly pressing needs to manage, control and channel the flux of visitors—yet must not ignore their duties to fulfil the missions of conservation and territorial development that they have been empowered with.

This need is shared by protected areas throughout Europe, a fact that prompted the Europarc Federation to draw up the methodology of the European Charter for Sustainable Development, generally known as the Charter.

Essentially, the Charter methodology coordinates and promotes the development of a model of sustainable tourism in European protected natural areas, which leads to the ratification by the Europarc Federation of the protected area in question as a Charter area.

8.3 Visitors to La Garrotxa Volcanic Zone

Currently, La Garrotxa Volcanic Zone Natural Park is visited annually by 400,000 people, mainly schoolchildren (60,000), families and guided tours. The busiest times of year are Easter, and the months of April, May, July, August, October and November.

In order to improve the quality of the visitor experience, the Natural Park has set up three information centres that attend over 90,000 people/year and 25 information points (i.e. tourist facilities with Natural Park accreditation). Additionally, it works closely with the guides of its accredited entities (who accompanied 15,000 visitors in 2014), has published a series of information leaflets and books, and has established a network of marked trails and itineraries that ensure that visitors can enjoy the main highlights of the Park's natural and cultural heritage without causing any disturbance to local residents (Fig. 8.3).

Fig. 8.2 Industry on the plain at Les Preses. *Credit* Llorenç Planagumà

To be able to manage the volume of visitors that flock to the area in spring and autumn, the Park's Special Management Plan states that transport and mobility in general must be adapted to the demands of tourism; this means better public transport, a plan aimed at providing access for people with reduced mobility, the promotion of 'soft' transport policies, and planning measures to improve and diversify the main areas of interest for visitors and equip them with appropriate infrastructures (e.g. car-parks and information points). Thus, walking, cycling, horse riding and the use of public transport have been encouraged by creating, adapting and signposting a network of paths and trails that both link places of interest for visitors and satisfy the demands of local walkers and hikers.

The correct location of car-parks is a key element in the management (Fig. 8.4) of visitor flows since they channel visits to particular sites. Furthermore, a well-planned network of car-parks can enhance the use of public transport, while the monitoring of their capacity is an excellent aid in the control of the intensity of visitor pressure. Information boards in car-parks also provide visitors with correct indications before they begin their visits, while accurate signposts encourage the use of the Park's network of marked footpaths and trails. All in all, the improvement of existing car-parks and the planning of the siting, capacity and infrastructures of new ones in accordance with the above-mentioned criteria are of vital importance in the correct management of public access to and use of the Natural Park.

The two main problems inherent in visitor use in La Garrotxa Volcanic Zone Natural Park are as follows:

(a) the existence of a non-sustainable development model at both local and national scales that negatively affects the conservation of the protected area. The Natural Park is a relatively densely populated area and much of its land surface is privately owned and devoted to intensive economic activities, all of which translates into serious pressure on the area that is esssentially incompatible with the correct conservation of its values;
(b) tourism based on short, low-quality visitor experiences centred both temporally and spatially on a few hyper-frequented sites, which provokes serious problems and can lead to a collapse in the quality of services and visitor experiences.

Fig. 8.3 Walkers on the Joan Maragall footpath in La Fageda. *Credit* Victor de Paz

8.4 Local Tourism Management

The management of tourism in La Garrotxa lies principally in the hands of Turisme Garrotxa, an association that acts as a nexus between the public and private sectors. It promotes tourism in the region and aims to improve the competitiveness of its members by offering information, guidance, assessment, training, promotion and product design and commercialisation. Tourism Garrotxa foments, represents, coordinates and participates in a model based on sustainable tourism that is locally well-rooted and combines respect for the environment with the needs and expectations of the entities, business and town councils that it assesses and promotes.

In 1994 a tourism advisory council was created in La Garrotxa, which—inspired by the French model of the *Pays d'Accueil Touristique*—was transformed two years later into Associació La Garrotxa Terra d'Acolliment Turístic and, subsequently, into the commercial brand 'Turisme Garrotxa'. Its objectives are countywide sustainability, quality, proximity and representativity, all in harmony with the protection of the natural environment. Its members include all the town councils in La Garrotxa, La Garrotxa County Council, the Consortium of L'Alta Garrotxa, La Garrotxa Volcanic Zone Natural Park, certain locally based associations (e.g. outward-bound centres, rural tourism, tourist activities, hostelries, camp-sites, rural accommodation, tourism and commerce in La Vall d'en Bas, walking companies and educational education entities such as Verd Volcànic), and the Centre for Territorial Sustainability. All the tourist-related businesses located in La Garrotxa that belong to one of Turisme Garrotxa's member associations are de facto 'Tourist Services' (unless they explicitly decide to opt out).

Fig. 8.4 Fageda d'en Jordà car-park. *Credit* Victor de Paz

All these businesses are Tourist Services and, unless their activities contravene the statutes of Turisme Garrotxa, require promotion; this means that, if their association is a member of Turisme Garrotxa, they can apply to be considered an 'Associated Tourist Service' and thus benefit from all the services offered by Turisme Garrotxa.

Turisme Garrotxa has four strategic objectives: to promote the environment, culture, society and tourist activities in La Garrotxa; to ensure that the catalogue of services offered by its members is kept up-to-date; to build mechanisms to guarantee coordination in the sector; and to redesign its own organisational structure, its staff and social projection. Since 2006, Tourism Garrotxa has fully embraced the European Charter for Sustainable Tourism model as a methodology for promoting sustainable tourism in the region.

8.5 The European Charter for Sustainable Tourism

The European Charter for Sustainable Tourism is an accreditation created and run by the Europarc Federation (Fig. 8.5), the body that acts as an umbrella for European public administrations and management bodies of natural protected areas (understood in a broad sense as national, natural and regional parks, biosphere reserves, and so forth). The Federation was founded in 1973 and currently embraces around 400 member organisations in 36 countries. Its task is to provide practical support for its members in the conservation of protected areas and to foment sustainable development in Europe based on a comprehensive, holistic vision of biodiversity and landscape management.

EUROPARC
FEDERATION

CARTA EUROPEA
DEL TURISMO SOSTENIBLE
EN LOS ESPACIOS PROTEGIDOS

Fig. 8.5 Logotip of the European Charter for sustainable tourism

The Federation is divided up into eight territorial sections: Atlantic Islands, Central and Eastern Europe, Germany, France, Italy-Federparchi, the Low Countries, Nordic-Baltic, and Spain. The task of protected areas is to control and organise the flow of tourists in order to guarantee the preservation of the values that they are intended to protect and to influence local territorial development. Thus, at the beginning of the 1990s the Europarc Federation created a working group entrusted with the task of drafting a proposal for working towards a balance between effective protection and tourism.

The result of this group's work was the report *Loving them to death. Sustainable Tourism in Europe's Nature and National Parks*, a document that recognised that tourism could have a negative effect on conservation in protected areas. It outlined a series of proposals for managing the double-edged sword of tourism 'in' and 'around' natural areas that tackled the key issues in sustainability requiring urgent resolution and recognised the need to work with those who operate in the tourist industry.

The efforts of this working group bore fruit at the end of the 1990s with the publication of a methodology, the *European Charter of Tourism in Protected Areas*. Among others, La Garrotxa Volcanic Zone Natural Park was tested as pilot region and the methodology was finally approved in 2000. According to the Europarc Federation, it has two chief objectives:

- Increase awareness of and support for Europe's protected areas as a fundamental part of our natural and cultural heritage, which should be preserved for and enjoyed by current and future generations;
- Improve the sustainable development and management of tourism in protected areas to ensure that the conservation of territorial values are compatible with satisfying the aspirations of businesses, the expectations of visitors and the needs of local residents.

Thus, this document is a practical management tool whose application, based on a voluntary agreement, aims to construct a model of sustainable tourist development in protected areas throughout Europe. The Charter consists of a credential awarded by the Federation that certifies that a region possesses a tourism model that (i) fulfils sustainable development criteria, (ii) has been developed in consensus with local stakeholders, and (iii) works in collaboration with the tourist sector.

At the end of the evaluation process, the Federation publically recognises the work that has been put in and awards the Charter to the protected area, which thus distinguishes the area in question as a 'sustainable tourist destination'.

8.5.1 The 'Knock-On' Effect of the Charter

To ensure that all the implicated parts in the value chain of tourism in the protected area participate in the move towards sustainability, the Charter is applied in different parts or phases that are defined by the area under evaluation.

- Part I: awarding of the Charter to the protected area. After an evaluation process involving the participation of stakeholders in tourism in the protected area, the area is awarded the Charter, which is valid for a period of five years. The same methodology is used throughout Europe and has been in use since 2000.
- Part II: engagement of local businesses with the Charter. In this stage, the park or other entity that has been awarded the Charter initiates a process whereby local tourist businesses commit themselves to the Charter after the drafting of a diagnosis, the implementation of a minimum of sustainable actions, and, finally, the definition of an action plan for continuous improvement. After three years, businesses are obliged to renew their action plans. Each Europarc territorial section draws up its own methodology (approved by the Federation) based on a common framework designed by the Federation. In the case of Spain, the methodology was approved in 2009 and the first stakeholders were accredited the following year.
- Part III: engagement of tour operators and travel agencies. This third stage is also drawn up by each territorial section and then approved by the Federation. In Spain, the methodology was finally approved at the end of December 2015 but its full content has not yet been disclosed.

In summary, the three stage of the Charter are as follows:

Part I: a territory is awarded the Charter based on the sustainability of its tourist strategy in and around the protected area in question.

Fig. 8.6 Businesses who have been awarded the Charter. *Credit* Octavi Bonet

Part II: local tourist businesses commit to the Charter and receive individual recognition of their engagement with and links to the protected area.
Part III: tour operators commit to the Charter through the creation of sustainable tourist products in collaboration with other businesses committed to the Charter and other services that respect the values of the natural protected area.

This cascade or 'knock-on' effect allows all types of local businesses (Fig. 8.6) and entities that come into contact with visitors (i.e. travel agents who sell a product to a tourist, a hotel, a guide or an information centre) to actively receive Charter ratification and participate in territorial tourism planning based on a model that combines conservation, local development and partnership in the tourist sector.

Part I of the Charter involves both the 'what' and the 'how', two key themes in sustainability in protected areas and in ensuring that consensus is reached in the territory. Part I of the Charter is thus an accreditation or recognition awarded by the Europarc Federation to protected areas (and, if need be, to their areas of influence) that is based on actions that fulfil the following requisites: "what" and "how".

'What', refers to the key topics and documents. The candidate territory, La Garrotxa Volcanic Zone Natural Park, must first undertake an assessment or diagnosis of its potential and prepare a Strategy and Action Plan that will satisfy the 10 topics—known as the 'Charter principles'—that Europarc defines as key areas and must be tackled when promoting sustainable tourism in an area. During the renewal process, which takes place every five years, the diagnosis is substituted by a report that analyses the progress made during the five years of the action plan that is coming to an end.

The ten Charter principles are as follows:

1. To involve all those implicated by tourism in and around the protected area in its development and management.
2. To prepare and implement a sustainable tourism strategy and action plan for the protected area.
3. To protect and enhance the area's natural and cultural heritage, for and through tourism, and to protect it from excessive tourism development.
4. To provide all visitors with a high-quality experience in all aspects of their visit.

5. To communicate effectively to visitors about the special qualities of the area.
6. To encourage specific tourism products which enable discovery and understanding of the area.
7. To increase knowledge of the protected area and sustainability issues amongst all those involved in tourism.
8. To ensure that tourism supports and does not reduce the quality of life of local residents.
9. To increase benefits from tourism to the local economy.
10. To monitor and influence visitor flows to reduce negative impacts.

In December 2005, the Europarc Federation published a revision of these principles, first debated in the 1990s and in force since 2001, that reformulated them into just five key areas:

- Give priority to protection
- Contribute to sustainable development
- Engage all stakeholders
- Plan sustainable tourism effectively
- Pursue continuous improvement.

The European Charter for Sustainable Tourism in La Garrotxa defines six general sustainable tourism objectives for the whole region: (i) To develop a quality policy, (ii) To organise and manage tourism, (iii) To promote, communicate and develop responsible marketing, (iv) To train, educate and raise awareness, (v) To protect and evaluate heritage (cultural, natural and geological), and (vi) To support the local economy and improve the quality of life of residents.

'How' refers to the process. Principle 1 and 15 years of work with accredited protected areas emphasise the importance of working with local stakeholders that have a special interest in tourist activities. These include the managers of the protected area, local bodies working in environmental protection, organisations that promote the local economy and tourism in particular, which in la Garrotxa have had the commitment and support of the following organisations: Turisme Garrotxa, La Garrotxa Regional Council, Garrotxa Líder Foundation and Olot Town Council. Businesses that In La Garrotxa have signed the Charter include landowners, neighbourhood associations in areas nearest to honey-pot sites, leisure groups that use the protected area and appreciate its inherent values, walkers, anglers, hunters, guides, farmers and school teachers and other educators.

These key agents, whose exact make-up will differ in each area, are known as the Permanent Forum, and their task is to participate in the diagnosis of the candidate area that takes into account the 10 Principles of Sustainable Tourism. These participants will also define a Strategy and Action Plan whose objective is to improve the territory's performance in relation to sustainable tourism. The Europarc Federation also upholds the importance of the work of the Permanent Forum as an effective framework for piloting the development of the Action Plan.

The extent to which these two principal blocks of requisites have been fulfilled is checked by a verification visit or audit in situ, which the Europarc Federation programmes once it receives all the documentation (in the case of both an initial application and a renovation), that is, the Diagnosis (or Report in case of a renewal) and the Strategy and Action Plan. The verifier's report produced by an appointed expert will be assessed by the Evaluation Committee, which will either grant, grant but with certain conditions, or deny the awarding of the Charter.

There are currently 143 accredited areas in Europe in 17 different countries: Bosnia and Herzegovina (1), Croatia (3), Denmark (1), Estonia (1), Finland (3), France (30), Germany (4), Italy (29), Latvia (1), Lithuania (2), the Netherlands (3), Norway (1), Portugal (6), Serbia (2), Slovakia (1), Spanish State (42) and Great Britain (13).

Of the 42 accredited areas in the Spanish State, four are in Catalonia:

- La Garrotxa Volcanic Zone Natural Park (Charter awarded in 2001 and renewed in 2006, 2011 and 2015)
- Ebro Delta Natural Park (Charter awarded in 2007 and renewed in 2011)
- Montseny Natural Park (Charter awarded in 2011 and renewed in 2015)
- Sant Llorenç del Munt i l'Obac Natural Park (Charter awarded in 2011 and renewed in 2015).

Part II of the Charter involves local tourism businesses wishing to benefit from the value of their surroundings. Apart from participating in the Permanent Forum, these businesses can take a step further in their commitment to the local tourism strategy by working towards the awarding of the Charter. This process is open to four types of businesses: accommodation, restaurants, activities providers (e.g. wildlife, cultural o walking guides) and food producers that also offer activities for visitors.

For businesses, commitment to the Charter has two objectives:

- To foster closer collaboration between protected areas that have been awarded the Charter and tourist businesses, based on mutual commitment to advance towards a more sustainable form of tourism.
- To distinguish those businesses that voluntarily commit themselves to working towards sustainable tourism in the

Charter area in which they operate, which will help guarantee continuous improvement in the sustainability of their businesses.

It is thus a process that recognises and awards businesses that have a true desire to improve their links with the protected area and engage in a strategy that will lead to greater sustainability both in the area in general and in their businesses in particular.

Beyond establishing the basic framework for action and ratifying that of each territorial section, the Europarc Federation does not intervene directly in Part II of the accreditation process. Thus, in this phase the dialogue always takes place between the candidate businesses and the protected areas.

To be eligible to opt for the Charter, tourist businesses must be located or carry out their activities in the Charter area; they should fulfil the demands of any relevant legislation at the time of certification; their activities must be compatible with a sustainable tourism strategy, the management plan of the protected area and the Charter sustainable tourism strategy; and they should join and participate actively in the Permanent Forum. In La Garrotxa Volcanic Zone Natural Park there are 17 of such tourist businesses in the La Garrotxa, including restaurants, rural tourism and ecotourism services, hotels and hostels, and local industries.

The certification process that businesses must undergo to be awarded the Charter consists of three key areas that are analogous to the 10 Charter principles:

- Improve the services on offer and links to the protected natural area
- Improve environmental performance
- Support local development and the conservation of local heritage.

The main phases of the certification process are also related to these three key areas:

- The initial diagnosis identifies the minimum actions to be carried out in each key area. These are known as the 'Key Actions' and all must be carried out within a set period of time. For instance, the company must draw up a list of the names and basic characteristics of its clients and send it to the protect area; it must attend training courses on tourism management, sustainable tourism or the natural protected area; it must create informative materials for its staff on subjects such as water, energy, waste, contamination and responsible consumption; and it must provide visitors with information on local shops, markets and products. Each protected area can also define other tasks that it regards as essential, which the candidate businesses have to fulfil together with the requirements of the Europarc section to which they belong.
- Verification visit: the verifier checks how the key areas are being implemented and outlines the content of the Action Plan.
- Definition of an individual Action Plan that should include at least three actions in each key area.
- Signing of a partnership agreement with the protected area. This agreement contains the list of actions that businesses agree to implement over the coming three years, as well as a commitment by the protected area to promote the tourism businesses in question. Although the commitment of the protected area with the candidate businesses are formulated generically at the beginning of the certification process, if need be the protected area may set out specific actions for each candidate business to fulfil.

The type of actions—as well as how they are to be fulfilled and monitored—contained in the agreements that the candidate businesses and the protected areas sign are generally highly diverse. Examples of commitments undertaken by protected areas include:

- Provision of publications and information about the protected area.
- Creation of new specific information material about the businesses and their inclusion on park websites.
- Funding of the costs of the certification process, calculated by the Spanish section of Europarc to be about 700€ per business.
- Priority given to the Charter-accredited businesses in activities organised by the protected area.

In terms of the impact of the certification process, the main positive results for businesses are as follows: an improvement in the relationship and communication with the protected area's management team (1.56 on a scale of 1–3); collaboration with other certified businesses (1.44); improvement in clients' implementation of measures designed to save and make more efficient use of resources, the fruit of the information that businesses learn during the certification process (1.95); and the support for local products and services associated with tourism (foodstuffs, crafts, etc.) (1.12).

Examples of the commitments by tourist businesses in La Garrotxa include:

- Creation of a specialised library.
- Passing on of information to the protected areas regarding any incidences (signposting, maintenance, litter, etc.) they detect.
- Elimination of non-reusable packaging in picnics.
- Naming of local producers on menus.

8.6 Main Impact of the Charter

The positive impact that all the regions that have implemented the Charter (Part I) agree upon is its ability to create a framework for stable working conditions that truly encourages mutual awareness between the stakeholders that operate, live and work in la Garrotxa area. Thus, despite the number of actions required to carry out effective coordinated actions, the Charter does provide a methodology whose requisites are clear and which facilitates collaboration between all interested parties.

Each area must decide on its own particular capacity to work within the Charter's framework and its ability to overcome the barriers that exist between different fields of work and administrative functions, and how it can guarantee true continuity to all the concerted work that is carried out.

The Charter also bestows a relevant role on the local population—a group that is often forgotten during any discussion of territorial development—in the definition of tourism strategies. Once again, each territory will decide whether the local population is sufficiently well represented by their elected municipal representatives, or whether local residents needed to be implicated more directly in the work of the protected area (Fig. 8.7).

In terms of content, both the Strategy and Action Plan (Part I) and the commitments undertaken by businesses

Fig. 8.7 Overview of Olot. *Credit* Marc Planagumà

Fig. 8.8 Rural accommodation. *Credit* Llorenç Planagumà

(Part II) tackle key aspects of sustainability. Consequently, the Charter also guarantees—as part of the road book that both the territory and businesses assume—the promotion of tourism-related activities that will have a positive impact on the territory, and improvements in the economic performance of local tourist businesses and in the quality of life of residents in the area of influence of La Garrotxa.

The continuing close collaboration between partners and the content of the Charter re-evaluation process established by the Europarc Federation both require annual monitoring. This exercise consists of mutual evaluation by all the partners of the Charter agreement based not on simple written reports (one area where we need to improve) but on project monitoring carried out through working meetings, the implementation of proposed actions, and so forth.

Finally, the commitment by businesses—the final phase to be implemented—implies the signing of agreements between these businesses and protected area management bodies, which represents a break with the tradition whereby the public administration grants accreditation and benefits to the private sector without assuming any reciprocal commitment (Fig. 8.8).

To date, Parts I and II of the Charter have been fulfilled in La Garrotxa, which have thrown their weight behind the process, above all in light of the benefits they have obtained from the development of common strategies and the implication of a greater variety of partners. The implementation of Part III and, thus, the implication of tour operators will represent a step forward in the commercialisation of tourist packets and a completion of the value chain of sustainable tourist activities.

9 Management Plan

Llorenç Planagumà

This chapter reviews the management of the geological heritage of La Garrotxa Volcanic Zone Natural Park, and includes both an evaluation of the strategy that was approved in 2004 and a discussion of the current and future work devoted to conserving the area's remarkable natural heritage and fomenting its sustainable use.

The geological heritage of the Natural Park contains two main elements, namely, the morphologies of its volcanoes and the individual outcrops and features. Its importance lies in the singularity of its materials, the characteristic features and forms that have been fashioned by volcanic activity, and the representativeness of the processes thus generated, which are the product of hundreds of thousands of years of geological activity. The best-conserved volcanic cones are protected as natural reserves, while individual geological outcrops are catalogued as 'outcrops of geological interest' (Fig. 9.1).

Two crucial issues greatly affect how the zone's geological heritage is managed: firstly, this volcanic landscape is also the physical foundation of other types of local heritage (i.e. forests, agriculture, industry, built-up areas, etc.) and, secondly, the Natural Park is an economically very developed and highly humanised area in which land is almost all privately owned. Thus, the management of the volcanic substrata is fraught with difficulties, above all because its preservation has to be compatible with the other uses that are made of it.

Declared in 1982 by a law passed by the Catalan Parliament, La Garrotxa Volcanic Zone Natural Park (15,000 ha) lies in the central part of the county of La Garrotxa and is home to a population of around 33,000 people. It protects 40 volcanic cones, of which 26 are natural reserves, and is the best-conserved volcanic landscape in the Iberian Peninsula. Built-up areas are excluded from the protected area and exist as islands within the park's boundaries (Fig. 9.1). In all, 98 % of the park's surface area is private property, and the presence of thousands of different landowners creates an added complication for the management of this singular protected area.

9.1 History of the Declaration of the Natural Park

The urban and industrial growth that La Garrotxa experienced in the 1970s seriously damaged the county's natural heritage. To stem this harm, sectors of local society began to mobilise and in 1976 the highly active Commission for the Protection of the Volcanic Zone was created. This mobilisation culminated a year later with the holding of the final act of the *Campaign to Save the Natural Heritage of the Catalan Counties*, organised by the Congress for Catalan Culture (Fig. 9.2).

Unanimously approved by the Catalan Parliament, Law 2/1982, 3 March, on the Protection of La Garrotxa Volcanic Zone, declared the area a 'Landscape of National Natural Interest'. In light of the singularity of the zone, its stated aims were the conservation of its flora, geomorphological features and particular scenic beauty (Article 1). This Law declared a series of 20 integral geobotanical reserves in which any action that might destroy, deteriorate, transform or disfigure the geomorphological outcrops and flora was prohibited (Article 2). Decree 71/1986, 13 February, defined the exact limits of the Natural Park and its natural reserves, and identified the ownership of all the land included in the natural reserves.

The first of the events that triggered the struggle to protect the volcanic area took place on 3 May 1966, the day that, based on a report by the laboratories of the Superior Council on Mining, the General Directorate of Mining granted an 861-ha concession—'Santa Margarida' (nº. 3140)—for the extraction of 'pumice stone' from the volcanoes Croscat, Santa

L. Planagumà (✉)
Tosca, Environmental Services, Olot, Spain
e-mail: llplanaguma@gmail.com

J. Martí and L. Planagumà (eds.), *La Garrotxa Volcanic Field of Northeast Spain*,
Geoheritage, Geoparks and Geotourism, DOI 10.1007/978-3-319-42080-6_9

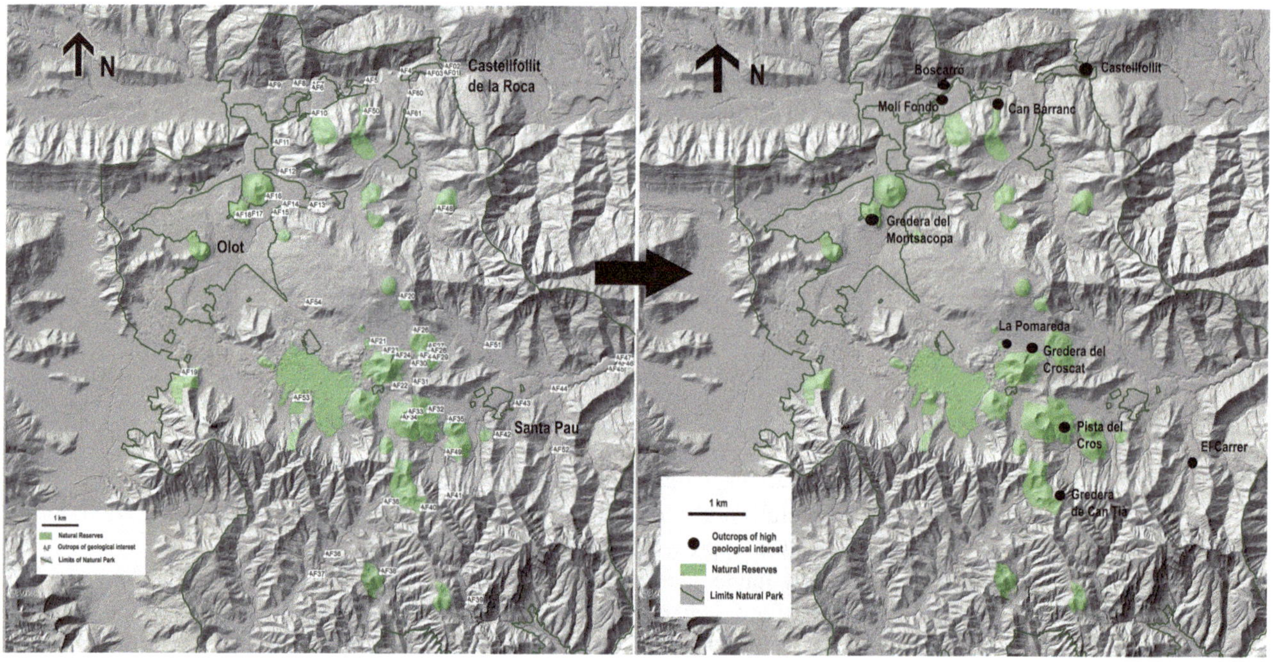

Fig. 9.1 Map of the outcrops of interest and selection the outcrops particular interest in La Garrotxa Volcanic Zone Natural Park

Margarida, Puig de Martinyà and Puig Astrol. By the beginning of the 1970s numerous projects—of which the most serious was the extraction of volcanic material (lapilli or *greda*) from the volcano Croscat and surrounding areas—were having a serious detrimental effect on the environment of the volcanic zone. Other threats to the local environment at the same time included the exploitation of radioactive minerals, the continued draining of sewage into the river Fluvià and its tributaries, the proliferation of fly-tipping (sometimes even at the foot of the area's most spectacular basalt outcrops such as Fontfreda in Sant Joan les Fonts), the construction of new farms and service facilities on the volcanic cones, and largely unplanned urban and industrial growth.

In 1975, the company Minas de Olot, SA was constituted from two pre-existing companies, Petrofísica Iberica, SA, from Madrid, and Lavas para la Construcción S.A. These two companies had been awarded adjoining concessions for mineral extraction in La Garrotxa but decided to create a single new company to put an end to the disputes that were constantly arising. Soon after, in 1976, a group of experts formed in Olot the Commission for the Protection of the Volcanic Zone (CPPZV), whose aim was to publish periodically information that would bring to light the rapidly increasing degradation of the volcanic area. Some authors consider that this organisation was the spark that detonated the protests and the struggle to protect the volcanic landscape in La Garrotxa. Its main aim was to promote the study of the volcanic zone and so it began to lobby scientists to support their demands for the declaration of a natural park (Fig. 9.3). This campaign, organised by local scientists, gained widespread support from numerous local socio-cultural and political groups. Nevertheless, on 27 May

Fig. 9.2 The quarry in the volcano Croscat when it was still being worked. *Credit* Pep Callís. *Source* Documentation Centre, Garrotxa Volcanic Zone Natural Park

Fig. 9.3 Demonstrations in favour of protecting the volcanoes. *Credit* Pep Callís. *Source* Documentation Centre, Garrotxa Volcanic Zone Natural Park

1977, the Directorate General of Mining and Construction of the Ministry of Industry granted the rights to the 'Santa Margarida' concession to Minas de Olot, SA for a period of 90 years. Yet, within five years, pressure groups had managed to persuade the Catalan government to pass Law 2/1982, 3 March, on the Protection of La Garrotxa Volcanic Zone, which declared the area a 'Landscape of National Natural Interest'. Subsequently, Law 12/85 on Natural Areas changed this original designation to 'natural park'.

Even so, the quarrying of the *greda* not was halted until the Catalan government acquired a majority shareholding in the mining company in 1990, nine long years after the original declaration of intent had been written into the first law protecting the volcanic area. This purchase meant that the Natural Park could now begin to restore the quarry in Croscat that, moreover, up to then had been used as the municipal rubbish tip. This was the first time in Catalonia that land had been expropriated to guarantee its conservation; in all, 77.2 ha were purchased and converted into the only natural reserve in La Garrotxa Volcanic Zone Natural Park that is fully publicly owned.

9.2 Regulations and Legal Framework for the Protection of the Geological Heritage

The legal framework that regulates the protection of the geological heritage of the area has been established by the following enactments (in chronological order): Law 2/1982, 3 March, for the Protection of La Garrotxa Volcanic Zone; Decree 113/1983, 12 April, that enforces Law 2/1982; Decree 41/2008, 26 February, that establishes the topographical boundaries of La Garrotxa Volcanic Zone Natural Park, its partial natural reserves and natural reserves; and government resolution 161/2010, 14 September, that approved the final version of the Special Management Plan of La Garrotxa Volcanic Zone Natural Park.

This series of laws, decrees and resolutions state that the natural reserves (i.e. all the best-preserved volcanic cones) are natural areas that warrant special protection, conservation and improvement in order to prevent any action that might lead to the deterioration, transformation or disfiguring of their geomorphology or flora.

The protection of the geological and edaphic heritage is covered by Article 1 of the park regulations detailed in the Special Management Plan, whose purpose is to protect the gea of the area. Amongst other considerations, this Special Plan contemplates the following regulations: no activities may be carried out that may provoke or accelerate erosive processes or increase the instability of the substrata; the singular volcanic soils whose extent is detailed in the maps included in reports must be strictly preserved. These soils are classified as of 'edaphological interest or as soils 'of agricultural interest' (Articles 41 and 42), other than when found in areas classified as of prime natural interest. The Natural Park's regulations guarantee the conservation and quality of these soils.

The natural reserves established by Law 2/1982 and Decree 41/2008, together with any other further volcanic edifices that are discovered and incorporated into the Natural Park in the future, are areas of prime natural interest, as stated in the regulations established in Chap. 7 of the Park regulations.

In and around outcrops considered of geological interest (Fig. 9.1), catalogued in Annex 1 of these regulations (shown on the diagrams of the visual units and on planning diagram 3 by the letters AIG), the following norms apply. No activity that might degrade or deteriorate the protected area or hamper its observation or study is permitted. Specifically, excavations and other earth movements are forbidden unless they are designed to improve or rehabilitate volcanic outcrops for scientific or educational purposes, and have been approved by the protected area's management team. The esplanades adjacent to the former quarries should not be built upon and should be kept free of installations that might obstruct the observation of the volcanic outcrop. The Natural Park should give priority to the active management of outcrops of exceptional interest and seek agreements and collaboration with landowners as a way of guaranteeing the maximum protection for these sites, if necessary in combination with external support. The Natural Park's Protection Council can decide to incorporate any new outcrops of interest that appear or are discovered into Annex 1 of these regulations. Any such incorporation will set in motion the procedures designed to modify the Special Management Plan, as well as the provisional but immediate application of the content of the previous sections as an interim measure. The Protection Council will commission an inventory of all the rootless cones situated in the area governed by the Special Management Plan, and the Natural Park's management team will perform the appropriate actions to conserve them.

9.3 Strategy for the Management of the Geological Heritage

The Natural Park's strategy for the conservation of the area's volcanic landscape stems from its management team's need to evaluate the current state of knowledge, conservation and awareness of what is one of the most valuable natural areas in Catalonia and the principal object of the protection work carried out in the Natural Park.

The drafting of the park's new Special Management Plan—the protected area's main tool for territorial and resource management (the first Plan was passed in 1994 and the current one in 2010)—took into account the need for a specific strategy for the management of the area's geological heritage (i.e. its vulcanism) based on three different spheres of action: research, conservation and awareness raising. The management strategy drawn up in 2010 highlighted the fact that, despite improvements in the geomorphological conservation status of the volcanic cones since quarrying activities were halted and restoration projects set in motion, in many sites (i) land management designed to highlight the morphology of cones and craters was still lacking, and (ii) the visibility of many important outcrops had deteriorated due to fly-tipping, landslips and vegetation growth. The ease of access to certain outcrops had also worsened. The strategy recognises that these tendencies are generally reversible and can be corrected by applying suitable measures.

The main recommendations for conserving the area's volcanic features are as follows: (i) publish a Special Management Plan for the volcano Montsacopa, (ii) implement a policy of land purchase and agreements with landowners in natural reserves, (iii) improve the visibility of the best-defined craters in the area (e.g. Roca Negra, Montsacopa, Puig Astrol and Croscat) (Fig. 9.4), (iv) restore and protect the pastures and arable land in craters and around the base of the volcanic edifices—for instance, Roca Negra (Fig. 9.4), Croscat and the crater of Llacunagra—to improve their visibility and geomorphological interpretation, as stated in the park's first Special Management Plan.

As part of the task of establishing a network of well-designed and properly maintained observation points, it will also be necessary to stabilise and conserve the walls and taluses of volcanic materials in the most significant outcrops in the area (Fig. 9.1), and to improve access to and the signposting of the pedagogically most interesting sites that have become partially concealed by vegetation growth or

Fig. 9.4 Volcanoes of Rocanegra and Santa Margarida. *Credit* Tosca, Environmental services

landslips. As well, it is essential to establish priorities for interventions based on didactic and vulnerability criteria, determine procedures for evaluating ephemeral outcrops exposed by public or private construction projects in the Natural Park, and improve the capacity to incorporate areas and outcrops of interest into municipal planning strategies.

9.3.1 Research in the Area

Research is regarded as an essential activity and one of its objectives is to provide the knowledge and information required for conserving the area's most important geological sites and improving awareness of the Natural Park and its heritage. In 2000, the main lacunae detected in volcanic research in La Garrotxa were in geochronology and the characterisation of the area's eruptive activity—more work is needed on some of the park's best-known volcanoes such as Santa Margarida and, for example, on how this volcano is related to other neighbouring volcanoes. At a secondary level of importance, a need for more work in fields such as tectonic structures and petrology was detected and, at thirdly, for more studies on palaeoclimatology, cartography and volcanic risk. As a result of this diagnosis, a series of measures under the umbrella of a single overall objective—the need to improve knowledge of vulcanism in La Garrotxa Volcanic Zone Natural Park—were proposed to redress these deficiencies.

Six specific objectives were defined:

1. **Create a digital database** to provide instant access to the state of volcanic research in La Garrotxa. Since then, the database of the project Vulcà has been set up based on a GIS.
2. **Characterise the eruptions of the area's most significant volcanoes**, above all Santa Margarida, Croscat, Montsacopa and Rocanegra. As Chapter two of this book reveals, studies giving new details of the eruptions of the volcanoes of Sant Margarida, La Garrinada and Montsacopa have recently been published.
3. **Promote collaboration with universities and other research centres** to finalise the task of dating the main volcanic features (i.e. the natural reserves and other most significant outcrops) in La Garrotxa. Of all the main objectives, this is the one that has advanced least, and the only collaboration to date that has got underway is with intern students from the Autonomous University of Barcelona. Ten years on, the park's geochronology and

the accurate dating of its volcanoes are the most obvious gaps in the scientific knowledge of the area.

4. **Find ways of integrating recently published studies on volcanic and seismic risk** into territorial management. In 2014 the first article on the potential volcanic hazard of the area was published. However, risk analysis is still lacking and has not yet been included in any management plans.
5. **Publish geological maps** at 1:25,000 scale of the Natural Park (Fig. 9.5) and its area of influence for fields such as lithology, tectonics, morphology, vulcanology and hydrology. Within the past 10 years, the Natural Park and the Catalan Institute of Cartography have published a vulcanological map of the area; subsequent revisions have led to the printing of a schematic vulcanological map of the area's eruptions (Figs. 9.6 and 9.7).
6. **Plan the monitoring of ephemeral outcrops** and incorporate the fresh information generated by research, construction work and boreholes into the database (see point 1). This objective has been fulfilled over the past 11 years and its results have provided important information for interpreting and conserving the volcanoes of the area. Over 100 sites have been inventoried and have afforded fresh data on, for example, known deposits, the boundaries of mapped features, stratigraphy and new volcanic edifices.

9.3.2 Conservation of the Volcanic Geological Heritage

The conservation status of the volcanic features of the area has been analysed using as a reference the Natural Park's revised Special Management Plan, the planning projects of the municipalities that possess one or more volcanic feature or outcrop, and the restoration projects and territorial plans that affect the outcrops included in the strategy. In general, the conservation status of the geomorphology of the volcanic cones has improved since quarrying was halted and restoration plans have been completed. Nevertheless, there are still a number of sites in which land-use management aimed at highlighting the morphology of cones and craters could be improved; however, most are privately owned and difficulties often arise when negotiating with landowners for permission to carry out such improvements. This means that

Fig. 9.5 Vall del Brugent through which the Amer Fault passes. *Credit* Llorenç Planagumà

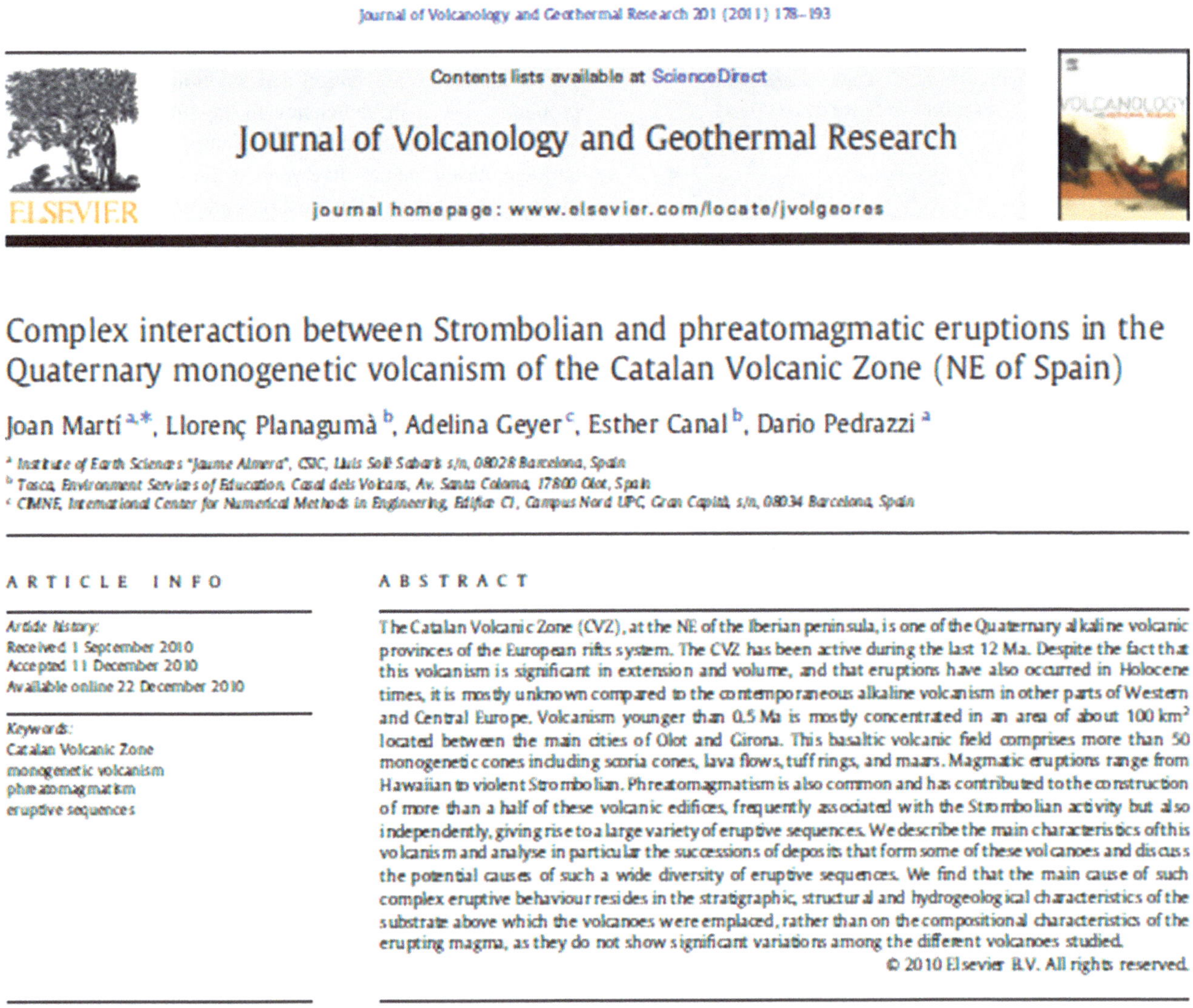

Journal of Volcanology and Geothermal Research 201 (2011) 178–193

Contents lists available at ScienceDirect

Journal of Volcanology and Geothermal Research

journal homepage: www.elsevier.com/locate/jvolgeores

Complex interaction between Strombolian and phreatomagmatic eruptions in the Quaternary monogenetic volcanism of the Catalan Volcanic Zone (NE of Spain)

Joan Martí [a,*], Llorenç Planagumà [b], Adelina Geyer [c], Esther Canal [b], Dario Pedrazzi [a]

[a] Institute of Earth Sciences "Jaume Almera", CSIC, Lluis Solé Sabarís s/n, 08028 Barcelona, Spain
[b] Tosca, Environment Services of Education, Casal dels Volcans, Av. Santa Coloma, 17800 Olot, Spain
[c] CIMNE, International Center for Numerical Methods in Engineering, Edifici C1, Campus Nord UPC, Gran Capità, s/n, 08034 Barcelona, Spain

ARTICLE INFO

Article history:
Received 1 September 2010
Accepted 11 December 2010
Available online 22 December 2010

Keywords:
Catalan Volcanic Zone
monogenetic volcanism
phreatomagmatism
eruptive sequences

ABSTRACT

The Catalan Volcanic Zone (CVZ), at the NE of the Iberian peninsula, is one of the Quaternary alkaline volcanic provinces of the European rifts system. The CVZ has been active during the last 12 Ma. Despite the fact that this volcanism is significant in extension and volume, and that eruptions have also occurred in Holocene times, it is mostly unknown compared to the contemporaneous alkaline volcanism in other parts of Western and Central Europe. Volcanism younger than 0.5 Ma is mostly concentrated in an area of about 100 km² located between the main cities of Olot and Girona. This basaltic volcanic field comprises more than 50 monogenetic cones including scoria cones, lava flows, tuff rings, and maars. Magmatic eruptions range from Hawaiian to violent Strombolian. Phreatomagmatism is also common and has contributed to the construction of more than a half of these volcanic edifices, frequently associated with the Strombolian activity but also independently, giving rise to a large variety of eruptive sequences. We describe the main characteristics of this volcanism and analyse in particular the successions of deposits that form some of these volcanoes and discuss the potential causes of such a wide diversity of eruptive sequences. We find that the main cause of such complex eruptive behaviour resides in the stratigraphic, structural and hydrogeological characteristics of the substrate above which the volcanoes were emplaced, rather than on the compositional characteristics of the erupting magma, as they do not show significant variations among the different volcanoes studied.

© 2010 Elsevier B.V. All rights reserved.

Fig. 9.6 Title page of a research article published on La Garrotxa volcanic zone. *Source* Elsevier

certain outcrops are less visible than they should be. Three specific objectives framed within a single general aim of conserving the geological and scenic values of the volcanic area have been defined:

1. Restore and conserve pasture and arable land in craters and at the foot of cones to ease the observation and interpretation of their geomorphological features. Over the years, this has been carried out in the craters of the volcanoes of El Montsacopa and Sant Margarida (Fig. 9.8).
2. Restore and conserve the best-preserved and most interesting outcrops via the creation of a network of sites or points of geological interest, all properly maintained to ensure good-quality views of the volcanic feature. A ranking of outcrops in order of importance has been carried out, and those of most interest have been incorporated into the Natural Park's Guide to Volcanism. In some cases, outcrops have been cleared and signposted and, for example, a number of quarries (e.g. on the flanks of Montsacopa) have been restored following criteria based on landscape considerations (Fig. 9.9).
3. Integrate the areas and outcrops of interest into municipal planning when towns' territorial plans are revised.

9.3.3 Awareness-Raising

Theoretically, the central element of any policy aimed at raising public awareness of the volcanic features of the

Fig. 9.7 Cover of the vulcanological map of the area. *Source* Documentation Centre, Garrotxa Volcanic Zone Natural Park

Natural Park is the visit to and discovery of the main sites of interest. It is thus essential to ensure that access to sites is straightforward and that their correct interpretation is possible, if necessary with additional information provided by information boards and small interpretation centres. However, as a general introduction to the area, the ideal type of facility is a museum or large visitor reception centre.

The depth of knowledge of vulcanology possessed by the people that live in and visit the volcanic area has been studied using an analysis of the didactic 'products' that transmit the key themes relating to vulcanism; an understanding of these concepts is essential for appreciating the values of the volcanic area and thus the need to protect it. The term 'products' is used in a generic sense to include all educational and information programmes, publications, activities, facilities and services.

This analysis was based on the criteria and objectives established in the environmental education programmes designed by the Natural Park's Environmental Education and Public Use Area. Environmental education is understood as a broad-based pedagogical process whose objective is to raise awareness amongst the general public of the environment via an understanding of issues and a willingness to participate in the prevention and solution of environmental problems. The main deficiency in the strategy for dissemination of the importance of the Natural Park's geological heritage during the past five years is the lack of a geological map; albeit not one of the most pressing requirements for the general population or for tourists, this type of map is an important resources for naturalists, technicians and scientists. Under the general concept of educating the public about the volcanic values of the Natural Park, four specific objectives were highlighted:

1. Revise existing educational material (programmes, activities, publications, facilities and services) to determine flaws in the way in which vulcanism is explained. This aim is currently being tackled as publications are revised and re-published and facilities updated with fresh data from research on the volcanoes of the area.
2. Design activities and resources (publications and facilities) whose goal is improved understanding of the area's volcanic activity amongst local residents and tourists. In recent years, a Guide to Vulcanism, a vulcanological map and a leaflet aimed at the general public have been published; still lacking are publications for children and a poster illustrating the volcanic activity that has taken place in the region.

The Guide to Vulcanism, an indispensable tool for getting to know the area, was published in 2000. It has three main sections: general vulcanism, vulcanism in Catalonia and a description of 12 local outcrops that help explain this vulcanism. The aim of this guide was to provide readers in a single publication with all the relevant information for interpreting the local volcanic activity during a visit to the area.

In 2008, La Garrotxa Volcanic Zone Natural Park in collaboration with the Catalan Geological Institute and Catalan Cartographical Institute published a new vulcanological map. This resource, which incorporates novel geological data generated by research in 1995–2008, is an essential tool for understanding the area's geological highlights.

3. Equip the network of sites of geological interest with appropriate facilities (i.e. signposting and interpretation tools) to teach the importance of these sites and the need for their conservation. Despite its recent restoration, the Museum of Volcanoes does not fully function as an

Fig. 9.8 The crater of the volcano El Montsacopa. *Credit* Pep Callís. *Source* Documentation Centre, Garrotxa Volcanic Zone Natural Park

Fig. 9.9 The restoration of the quarries on the flanks of the volcano El Montsacopa. *Credit* Tosca, Environmental services

information point for visitors wishing to study the area's geology. Any such museum should provoke an interest in discovering volcanic sites and should act as a facility that complements the knowledge acquired by direct observation and interpretation in the field. Points of geological interest should fulfil the following criteria:

- be publicly owned or used with the landowner's permission.
- be adequately signposted from car-parks.
- be well conserved and correctly maintained.
- be large enough to be visited by a group (e.g. a class of children).
- visitors must have access to information about the site (publications or in situ).
- visits to the site should not negatively affect any other natural or social element.
- sites must not be vulnerable to damage.

4. Train guides to improve their knowledge of local vulcanism and ensure that up-to-date information generated by research reaches them. Each year the Natural Park, Turisme Garrotxa and the Olot Foundation for Further Education organise a training programme aimed at improving knowledge of La Garrotxa in the tourist sector (Fig. 9.10). This programme is part of the strategy of the European Charter for Sustainable Tourism in La Garrotxa. Although open to all, these courses target above all professionals in the tourist sector. Their general aim is to improve the quality of the tourist services offered in the county through courses that range in subject matter from general background knowledge of the local area to more specific tuition on subjects of particular relevance in La Garrotxa.

On 21–25 May 2012 Olot hosted VOLCANDPARK, the First International Congress on Management and Awareness-raising in Protected Volcanic Landscapes. This congress filled an important gap, since hitherto there had been no regular forum anywhere in the world in which the managers of protected volcanic areas could exchange knowledge and experience, or attempt to set up joint projects. The motivation behind the organisation of this congress was the desire to describe—modestly but compellingly—the work carried out to date in the Natural Park or, in other words, to discuss its successes and failures, and to establish a framework for the exchange of experiences with management teams from other protected volcanic areas. VOLCANDPARK was a great

Fig. 9.10 One of the sessions that form part of the training course for Natural Park guides. *Credit* Xavier de Bolós

opportunity for the Natural Park since it provided a forum for its management team to explain how the park is run; as well, it also helped put Olot and the volcanic zone of La Garrotxa on the map and make this volcanic region better known throughout the world.

9.3.4 Evaluation of 10 Years of Management

A detailed analysis reveals that the original objectives of research, conservation and awareness raising in the Natural Park have been almost completely fulfilled. In terms of research, of the 12 volcanoes that require further study, four have been well characterised, the stratigraphy of the volcanic zone is now much better understood, and almost all the planned maps have been published.

Conservation has continued, above all through the uninterrupted monitoring of ephemeral outcrops, which has generated new information on the Begudà and Castellfollit lava flows, and the volcanoes of La Garrinada, Pla de Massandell, Montsacopa and Puig Roser. The morphology of the volcanoes is today more visible than in the 1980s and their overall conservation state has remained unchanged over the past 10 years. Tourists, schoolchildren and scientists are able to observe and study the same geological processes as before, indicating that there has been no significant loss in the visitability of the outcrops.

Public awareness of the volcanoes has been boosted by the work of the local town councils, who have become involved in the conservation of the region's geological heritage. Local residents now have better knowledge of 'their' volcanoes and so the Natural Park's awareness-raising strategy should be regarded as generally positive, albeit with some room for improvement.

9.4 Future Action Plans

A number of strategic lines of work aimed at preserving the region's geological heritage, based on the general objectives set out by La Garrotxa Volcanic Zone Natural Park in its management plans, have been defined as follows.

9.4.1 Modification of the Sphere of Influence

1. Broaden the strategy and, therefore, the coordinated management of other nearby sectors such as La Vall de la Llémena, Crosa de Sant Dalmai and Camp dels Ninots that also possess significant volcanic features. Although there is no need to increase the surface area of the Natural Park, effective coordinated mechanisms are required for actions such as the conservation of existing outcrops and the monitoring of ephemeral outcrops (Fig. 9.11).

Fig. 9.11 Volcano of La Crosa de Sant Dalmai. *Credit* Xavier de Bolós

9.4.2 Research

2. Reach agreements with universities and other research institutes to carry out campaigns aimed at improving knowledge of the chronology of the volcanoes as a means of confirming the stratigraphy of the region.
3. Improve the database—and make it accessible to a wider public—that the Natural Park's GIS is currently using by incorporating and codifying all new geological information. This process should promote several citizen science programmes (e.g. the project to inventory the park's rootless volcanoes) and encourage public participation in scientific research and monitoring.
4. Promote petrological studies and the characterisation of volcanic edifices to fill in gaps in current knowledge in these fields (Fig. 9.12).

Fig. 9.12 Front cover of the Natural Park's schools programme. *Source* Documentation Centre, Garrotxa Volcanic Zone Natural Park

Fig. 9.13 Overview of the area to be conserved: landscape, volcanoes, outcrops, people and peace and quiet. *Credit* Marc Planagumà

Table 9.1 Table of proposed indicators for analysing future strategies

Social indicators	**Observations**
Average age of professionals who work in jobs related to the conservation of the volcanic heritage	One of the quality objectives is workplace stability; it is important to establish the average age of people working in jobs related to research, conservation and education in the volcanic area (i.e. Natural Park staff, guides, educators, professionals, etc.)
Satisfaction of workers in jobs related to the conservation of the volcanic heritage	Satisfaction of people employed in management: guides, educators, technical staff, professionals, etc.
Visitor satisfaction	Carry out a visitor satisfaction survey every two years
Environmental indicators	**Observations**
Geological processes for study and for observation are conserved over time	Determine which geological processes can be observed, and analyse if there has been any losses of over the years due to landslips, erosion, construction work, etc.
The number of scientific articles presented for publication annually	Articles that discuss the area's gea and landscape
Hours devoted annually to environmental education	Determine how many hours visitors devote to environmental education
Analysis of the state of conversation of the landscape	Analyse via photographs changes in land use in geological outcrops and volcanic cones
% of outcrops of interest that are adequately managed	Calculate how many outcrops are effectively managed either via stewardship agreements or the direct management of publicly owned land
Economic indicators	**Observations**
Study of the socio-economic impact (every four years)	The socio-economic impact of a protected area is, according to a recent study, 8€ for every 1€ invested. This figure should be revised every four years
Analysis of the resources devoted to research/conservation/education and awareness-raising	Devise a table to rank the resources devoted to each aspect in the Natural Park. Although there is no need for resource use to be equal, imbalances must be objectively justifiable. Education is a vital tool in conservation and in the coming years must be promoted and supported accordingly
Networking and participation	
Actions carried out in collaboration with local entities	The number of actions that are undertaken conjointly with local town and city councils
Actions carried out involving local entities, businesses and trade unions	The number of management actions in which local social and economic bodies are implicated

5. Bolster the monitoring of ephemeral outcrops by improving coordination with other administrative bodies and by broadening the range of such outcrops studied.

9.4.3 Conservation

6. Continue restoring the network of outcrops of interest via stewardship agreements or outright purchase. This type of conservation must include signposting, restoration and the regular clearing of outcrops.
7. Improve the visibility of the volcanoes so that both cones and craters can be properly appreciated. This is the case above all in the volcanoes of Rocanegra, Santa Margarida, Montsacopa and Croscat, as well as in other less visited volcanoes such as Can Tià and El Racó.
8. Promote knowledge of the geodiversity of the area beyond the region's already well-known volcanic geological heritage and include, for instance, sedimentary rocks dating from the Eocene, the landscape in general, and tectonic structures.

9.4.4 Awareness Raising

9. Create a plan for phasing out informative material, dossiers and other didactic material that has become out-of-date and replace it with material based on fresh knowledge of the area's vulcanism.
10. Endorse popular participation and citizen science as a tool for educating and raising awareness of important geological sites.
11. Support environmental education as the basis of conservation, and strengthen and improve existing education programmes. Re-launch the programmes *Escola i Entorn* ('School and Surroundings') and *Visc entre Volcans* ('I live amongst volcanoes').

9.4.5 Resources and Evaluation

12. Promote workplace job stability and guarantee better working conditions for people working in the area as an essential part of the conservation of its geological heritage. Move towards high-quality sustainable tourism. Continuous evaluation of the fulfilment of objectives via indicators will help implement the park's management plan (Fig. 9.13, Table 9.1).

9.4.6 Communication

A good communication policy is key for ensuring transparency and the participation of local residents. This can be achieved via a website, bulletins, articles in the local press and the holding of themed workshops. Programmed excursions to sites of geological interest should also be organised on a regular basis.

Recommended Bibliography and Sources of Information

Barde-Cabusson, S., Bolós, X., Pedrazzi, D., Lovera, R., Serra, G., Martí, J., et al. (2013). Electric resistivity tomography revealing the internal structure of monogenetic volcanoes. *Geophysical Research Journal,40*, 1–6. doi:10.1002/grl.50538

Barde-Cabusson, S., Gottsman, J., Martí, J., Bolós, X., Camacho, A. G., Geyer, A., et al. (2014). Structural control of monogenetic volcanism in the Garrotxa volcanic field (Northeastern Spain) from gravity and self-potential measurements. *Bulletin of Volcanology,76*, 788.

Bolós, A. (1925). L'estacio paleontolo`gica del Pont de Ferro i algunes consideracions sobre el vulcanisme olotı´. But. ICHN 2a se`rie. V, 4. Barcelona.

Bolós, X., Barde-Cabusson, S., Pedrazzi, D., Martí, J., Casas, A., Himi, M., et al. (2012). Investigation of the inner structure of La Crosa de Sant Dalmai maar (Catalan Volcanic Zone, Spain). *Journal of Volcanology and Geothermal Research,247–248*, 37–48.

Bolós, X., Barde-Cabusson, S., Pedrazzi, D., Martí, J., Casas, A., Lovera, R., et al. (2014). Subsurface geology in the monogenetic La Garrotxa Volcanic Field (NE Iberian Peninsula). *International Journal of Earth Science. 130*. doi:10.1007/s00531-014-1044-3

Bolós, X., Martí, J., Becerril, L., Planagomà, L. L., Grosse, P., & Barde-Cabusson, S. (in press). Volcano-structural analysis of La Garrotxa Volcanic Field (NE Iberia): Implications on the plumbing system. Tectonophysics.

Bolós, X., Planagumà, L., & Martí, J. (2014). Volcanic stratigraphy and evolution of the quaternary monogenetic volcanism in the Catalan Volcanic Zone (NE Spain). *Journal of Quaternary Science, 29*(6), 547–560.

Bonet, O. (2000). L'Educació ambiental als espais protegits de les comarques gironines in Dlleure: un espai al servei de l'educació. Girona: [s.n.], pp. 4–13

Bonet, O., Oliver, X., & Llongarriu, M. (1999). La Educación ambiental formal en el Parque Natural de la Zona Volcánica de la Garrotxa = Official environmental education in Natural Park of the volcanic zone of La Garrotxa in Enseñanza de las ciencias de la tierra : Revista de la Asociación Española para la Enseñanza de las Ciencias de la Tierra. Madrid: l'associació, Vol. 7, núm. 3, pp. 278–284.

Burjachs, F. (1985). Evolucio de la vegetacio i paleoclimatologia des de fa me´s de 85.000 anys a la regio d'Olot. *Ana`lisi pol lı´nica del Pla de l'Estany (Sant Joan les Fonts, la Garrotxa). Technical report. Vitrina: publicacio del Museu Comarcal de la Garrotxa, a Recerca Científica del Parc Natural de la Zona Volca`nica de la Garrotxa.* (Unpublished technical report. Consultable in Museum of La Garrotxa.) 1982–1992.Xxxxx1: 49–56

Cebriá, J. M., López-Ruiz, J., Doblas, M., Oyarzun, R., Hertogen, J., & Benito, R. (2000). Geochemistry of the quaternary alkali basalts of Garrotxa (NE Volcanic Province, Spain): A case of double enrichment of the mantle lithosphere. *Journal of Volcanology and Geothermal Research,112*, 175–187.

Cimarelli, C., Di Tragia, F., de Rita, D., Gimeno Torrente, D., & Fernandez Turiel, J. L. (2013). Space time evolution of monogenetic volcanism in the mafic Garrotxa Volcanic Field (NE Iberian Peninsula). *Bulletin of Volcanology,75*, 758.

Council of Europe. (2000). *European landscape convention*. Strasbourg: Council of Europe.

Council of Europe, Recommendation CM/Rec. (2008). *3 of the committee of ministers to member states on the guidelines for the implementation of the European Landscape Convention*. Strasbourg: Council of Europe.

Decret 343/2006. (2006). de 19 de setembre, pel qual es desenvolupa la Llei 8/2005, de 8 de juny, de protecció, gestió i ordenació del paisatge, i es regulen els estudis i informes d'impacte i integració paisatgístic, Diari Oficial de la Generalitat de Catalunya, 4723, pp. 39.384–39.388.

Donville, B. (1973). *Géologie Néogène et ages des éruptions volcaniques de la Catalogne orientale*. These doct. Fac. Sciences, Université Paul Sabatier, Tolosa de Llenguadoc.

Fanlo Grasa, E. (1995). Recerca sobre els continguts en matèria d'educació ambiental a la població local de la zona volcànica de la Garrotxa, PNZVG, 1995. 33f.+2 appendices. PNZVG grant awarded in 1994.

Fernández, M., Torné, M., & Zeyen, H. (1990). Lithospheric thermal structure of NE Spain and the North-Balearic basin. *Journal of Geodynamics,12*, 253–267.

Flecha, R., Padrós, M., & Puigdellívol, I. (2003). Comunidades de aprendizaje: transformar la organización esclar al Servicio de la Comunidad. Organización y gestión educativa (no 5, pp. 4–8). Bilbao: Fórum Europeo de Administradores de la Educación y CISSPRAXIS;SA.

Folch, R. (1988). Natura, ús o abús? Llibre blanc de la gestió de la natura als Països Catalans, 2ª edición, ICHN, Editorial Barcino, pp. 291–292.

Gallart, J., Pous, J., Boix, F., & Hirn, A. (1991). Geophysical constraints on the structure of the Olot Volcanic Area, north-eastern Iberian Peninsula. *Journal of Volcanology and Geothermal Research,47*, 33–44.

Graells, J. (2008). Algunes reflexions sobre competències bàsiques Cinquena Jornada d'ensenyament de les matemàtiques organitzada per FEEMCAT, XEIX I SCM. Barcelona, September 20, 2008

Grau, S. (1996). La protecció dels espais volcànics. Revista de Girona. 174, 85–89.

Guérin, G., Behamoun, G., & Mallarach, J. M. (1985). Un exemple de fusió parcial en medi continental. *El vulcanisme quaternari de la Garrotxa. Publicació del Museu Comarcal de la Garrotxa, Vitrina, 1*, 19–26.

J. Martí and L. Planagumà (eds.), *La Garrotxa Volcanic Field of Northeast Spain*, Geoheritage, Geoparks and Geotourism, DOI 10.1007/978-3-319-42080-6

Guerin, G., & Valladas, G. (1980). Thermoluminescence dating of volcanic plagioclases. *Nature 286*, 697–699.

http://www.deseco.admin.ch/bfs/deseco/en/index/03/02.parsys.78532.downloadList.94248.DownloadFile.tmp/2005.dscexecutivesummary.sp.pdf

http://parcsnaturals.gencat.cat/ca/garrotxa

http://www.redeuroparc.org/cartaeuropeaturismosostenible.jsp

http://european-charter.org/home/

http://territori.gencat.cat/web/.content/home/01_departament/estadistica/estadistiques_i_indicadors_de_medi_ambient/publicacions_estadistiques/documents/opusclecatala2012.pdf. Accessed on February 20, 2015.

http://territori.gencat.cat/web/.content/home/01_departament/estadistica/estadistiques_i_indicadors_de_medi_ambient/publicacions_estadistiques/documents/opusclecatala2013ind.pdf. Accessed on February 20, 2015.

http://www.redeuroparc.org/img/publicaciones/Impacto-FaseII-CETS-EUROPARC.pdf. Accessed on February 20, 2015.

Lewis, C. J., Vergés, J., & Marzo, M. (2000). High mountains in a zone of extended crust: Insights into the Neogene-Quaternary topographic development of northeastern Iberia. *Tectonics*, *19*(1), 86–102.

Llei 8/2005 i Reglament de protecció, gestió i ordenació del paisatge. Barcelona: Generalitat de Catalunya, Departament de Política Territorial i Obres Públiques. (Quaderns de Legislació).

Mallarach, J. M. (1982). Carta geologica de la regió volcanica d'Olot. Litologia i geomorfologia. 1120.000. Ed Maber, Ajuntament d'Olot.

Mallarach, J. M. (1992). Informe intern sobre la conservació de les àrees de reserva establertes per la llei 2/1982 de protecció de la Zona Volcànica de la Garrotxa. Unpublished document.

Mallarach, J. M. (1996). La gestió del patrimoni geològic, el cas de la Garrotxa. Revista de Girona, 174, 90–93.

Mallarach, J. M. (1998). El vulcanisme prehistòric de Catalunya. Diputació de Girona.

Mallarach, J. M., & Riera, M. (1981). Els volcans olotins i el seu paisatge. Barcelona: Editorial Serpa, 250 p.

Martí, J., Mitjavila, J., Roca, E., & Aparicio, A. (1992). Cenozoic magmatism of the Valencia trough (Western Mediterranean): Relation between structural evolution and Volcanism. *Tectonophysics,203*, 145–166.

Martí, J., et al. (2000). El Vulcanisme: guia de camp de la zona volcànica de la Garrotxa. Olot: PNZVG, 104 p.

Martí, J., Planagumà, L., Geyer, A., Canal, E., & Pedrazzi, D. (2011). Complex interaction between Strombolian and phreatomagmatic eruptions in the Quaternary monogenetic volcanism of the Catalan Volcanic Zone (NE of Spain). *Journal of Volcanology and Geothermal Research,201*, 178–193.

Ministerio de Obras Públicas, Transportes y Medio Ambiente. (1996). El PATRIMONIO GEOLÓGICO. Bases para su valoración, protección, conservación i utilización. Serie monografías.

Neumann, E. R., Martí, J., Mitjavila, J., & Wulff-Pedersen, E. (1999). Origin and implications of mafic xenoliths associated with Cenozoic extension-related volcanism in the València trough, NE Spain. *Mineralogy and Petrology,65*, 113–139.

Nogué, J., & Sala, P. (2006). Prototipus de catàleg de paisatge. Bases conceptuals, metodològiques i procedimentals per elaborar els catàlegs de paisatge de Catalunya, Olot, Observatori del Paisatge de Catalunya.

Nogué, J., & Sala, P. (coord.). (2010). Catàleg de paisatge de Comarques Gironines. Olot: Observatori de Paisatge; Barcelona: Departament de Territori i Sostenibilitat de la Generalitat de Catalunya; unpublished document.

Oliver, X. (1997). Programes de gestió de l'àrea d'educació ambiental, divulgació i ús públic del Parc Natural de la Zona Volcànica de la Garrotxa, 1997.7f.

Oliver, X. (2002). El Parc Natural de la Zona Volcànica de la Garrotxa. Olot: Llibres de Batet.

Olivar Fenrández de Villarán, M. (1995). Estudi de l'ús escolar al Parc Natural de la Zona Volcànica de la Garrotxa, PNZVG, 1995.104 pp. +9 appendices. PNZVG grant awarded in 1994.

Pedrazzi, D., Bolós, X., & Martí, J. (2014). Eruptive and depositional history of La Crosa de Sant Dalmai maar (Catalan Volcanic Zone, NE Spain). *Geosphere, 10*(1). doi:10.1130/GES00959.1

Plan especial de la Zona Volcánica de la Garrotxa. (1994). Departament de Medi Ambient.

Rychen. D. S., & Salganik, L. H. (eds.) (2003). Key competencies for a successful life and a well-functioning society, (Spanish translation: Las competencials clave para el bienestar personal, económico y social (1st ed. in Spanish, 2006)

Salganik, L. H., Rychen, D. S., Moser, U. & Konstant, J. (1999). Projects on competencies in the OECD context: Analysis of theoretical and conceptual foundations (Spanish translation: Projectos sobre competencias en el contexto de la OECD: análisis de fundaciones teoréticas y conceptuales (1st ed. in Spanish, 2000)

Wilson, M., & Downes, H. (1991). Tertiary-Quaternary extension-related alkaline magmatism in Western and Central Europe. *Journal of Petrology,32*, 811–849.

www.OECD.org/edu/statistics/deseco

www.deseco.admin.ch

Ziegler, P. A. (1992). European Cenozoic rift system. In: P. A. Ziegler (ed.), Geodynamics of rifting, vol. I. Case history studies on rift: Europe and Asia. *Tectonophysics, 208*, 91–111.

Kolibri 360 GmbH
Ferdinand-Jühlke-Straße 7
99095 Erfurt, Deutschland
produktsicherheit@kolibri360.de